The Last Days of
BRITISH
STEAM
RAILWAYS

The Last Days of
BRITISH STEAM RAILWAYS

COLIN GARRATT

PSL

Patrick Stephens, Wellingborough

First published in 1985

British Library Cataloguing in Publication Data

Garratt, Colin
The Last Days of British Steam Railways
1. Locomotives—Great Britain—History—
20th century—Pictorial works
I. Title
625.2'61'0941   TJ603.4.G7

ISBN 0-85059-781-1

Title Page *How one mourns for the days of healthy
goods yards and a railway which handled the nation's
freight. The Stanier '8F' 2-8-0s—of which there were
almost 700—were but one of numerous freight classes.
Here, No 48637—one of Toton's 50 '8F's—prepares a
train at Leicester North sidings.*

*Patrick Stephens Limited is part of the
Thorsons Publishing Group*

Text photoset in 10 on 11 pt Garamond by Avocet
Marketing Services, Aylesbury, Bucks. Printed in Great
Britain, on 115 gsm Clandon Matt coated cartridge,
and bound by The Garden City Press, Letchworth, Herts,
for the publishers, Patrick Stephens Limited, Denington
Estate, Wellingborough, Northants, NN8 2QD, England.

Contents

Introduction

'Plain and simple' were the words I used to describe this collection of monochrome prints to the publisher. They were pictures made when I was on my first foot as a photographer and a last ditch attempt to have some personal record of an age which had meant so much to me. Most were made during the mid 1960s, for although I had been closely associated with railways for 15 years before this, I had made the common mistake of taking few photographs — after all steam trains were going to last forever.

The great passion of those early years was train spotting and from 1949 I marvelled at Britain's incredibly rich locomotive history. This was a time when Britain's railways provided an efficient, comprehensive, safe and reliable national transport system. The bulk of the nation's freight and indeed a considerable percentage of passenger journeys were by rail and most trains were steam-hauled by locomotives fired with our own coal. As a transport network it was without parallel; communications were rapid, industries were sited within easy reach of the railway and there was little need either for private cars or juggernaut trucks. Railways are undeniably the most civilised form of land transportation and their destruction over the last 30 years will almost certainly be seen by future generations as one of the most short-sighted acts of the twentieth century.

Over the years following World War 2, Great Britain had some 30,000 steam locomotives in active service, embracing hundreds of different types, and my lineside vigils by that childhood bridge at Newton Harcourt initiated me into the sheer magnificence of a properly utilised railway. The aim of train spotting was to see all the members of every different class and when a class was completed it was said to be 'cleared'. Some classes consisted of only one or two engines, whilst others ran into hundreds, such as the 842 members of the LMS 'Black 5', examples of which could be seen in most parts of the country from Inverness to Bournemouth. Withdrawn engines not seen were recorded as lost. In railway works or even on scrap lines where locomotives were likely to be found in pieces, frames were countable but interchangeable items like boilers and tenders were not.

For millions, train spotting offered colour, romance and excitement; its greatness far surpassing superficial appearances. With so many locomotives active it offered sport indeed; it opened young eyes to the geography of one's country and gave incentive to travel; it revealed industrial history, developed the eye for detail and aesthetics, the mind for numbers and statistics, and it broadened comprehension of distribution and purpose; cause and effect. He who understood railways had his finger on the nation's pulse.

Train spotting trips to far away places — often at very young ages — will live on in the minds of those privileged enough to have experienced them and it is amazing how similar are the sentiments and emotions expressed today when individuals recall those golden years. The excitement of these trips contrasted with the thrill of the unexpected at our local lineside and certain daily workings brought a possibility of the rare and exotic to otherwise familiar locations. To us in Leicester, certain trains became almost legendary and were the subject of daily observation and comment — trains known to us by such terms as the 'Kingmoor', the '5A', the 'Edge Hill', the 'York Goods', the 'Western' and so many more. 'Shoppers' — engines either en route to or returning from major overhauls — also brought innumerable thrills.

During summer time the familiar train spotting places on main lines would attract hundreds of people; grass was worn off the embankment and we sat on hard dusty patches — a grandstand to one of the finest unfolding dramas of all time. As in great sport, the thrill of the unexpected loomed behind every quiet moment; 30 classes might be seen on one day and from these any rarity could pass. I remember once when a rare Scottish 'Jubilee' complete with huge St Rollocks numbers on her cab worked southwards through Rugby; 150 enthusiasts cheered wildly from the tracksides; pens, notebooks and sandwiches flew into the air amid uninhibited enthusiasm.

I did do a little intermittent photography as early as 1952 but the negatives have tragically been lost. During the years that the pictures in this book were made I was in a conventional marketing career and had no aspiration to have the work published; the pictures were, as I said earlier, only for my personal album to complement the wonderful memories I had. They were a last testament for I had no idea of what lay ahead for me and my over-riding mood at the time, often to the point of morbidity, was that the two things that I most cared about — steam railways and New Orleans jazz — were on the verge of extinction. The great but aged veteran jazz men of New Orleans were dying throughout the 1960s, almost as quickly as the familiar steam classes were disappearing and the only way that either steam railways or New Orleans jazz could survive would be as tourist revivals with all the inevitable anaesthetism which goes with them.

Little did I realise that once the trauma of steam's extinction in Britain was over, new perspectives would emerge; for not only did most of the railways' romance and atmosphere crumble with the dying steam engine,

but a vast reduction in track miles was undertaken; main lines, cross country lines, branch lines, sidings and marshalling yards were abandoned. Engine sheds suffered mass closure (there had been as many as 500 in 1950!) as a road-based economy began to take over — a trend which continues unchecked to this day.

Such decline was not restricted to Britain. All over the world historic locomotives and whole railway networks were under threat and exactly one year after British Railways' last steam train ran in August 1968, I abandoned my promising commercial career to professionally document the last steam locomotives of the world — but that is another story.

And so the pictures between these covers were from my apprenticeship as I blindly groped in an untutored way towards lucid photographic expression. It was a groping that would stand me in good stead over the unimaginable future which lay ahead, for never in my wildest dreams did I visualise that my career was destined to be with steam locomotives, for on that sunny afternoon in August 1968 when BR dropped the last fire to extinguish the industrial revolution's brightest light, I imagined that the happiest part of my life had come to an end.

Intermittently mentioned throughout this volume will be the companions who shared these monochrome years with me; Brian Stafford, David 'Don' Holland, George Brunavs and Judy Maddock. Brian and George also played in Colin Garratt's Superior Jazz Band, Don managed us for a time whilst Judy sang and was without doubt one of the finest Blues' artists Britain has produced.

So here, 'plain and simple', are some pictures and thoughts in the form of a brief personal evocation of happy times gone by.

Colin Garratt
Newton Harcourt, Leicestershire
March, 1985

Shrouds of steam disperse to reveal the friendly face of Ivatt 2-6-2T No 41299 bursting from beneath the portals of Eastleigh shed. Though of LMS origin, considerable numbers of these general purpose engines were allocated to the Southern.

Chapter 1

Scotland

My earliest awareness of Scottish railways was seeing photographs of the Highland Railway's 'Ben' Class No 54398 *Ben Alder* in my ABC stock book. I remember sitting by the line at Newton Harcourt — never having travelled — dreaming about this handsome engine which lingered in the far distant Highlands. Little less inspiring were the Great North of Scotland Railway 'D40's one of the most handsome inside-cylinder 4-4-0s of all time.

Over the next few years I came to regard Scotland with awe; its railway history had fallen into perspective and my keenness to see celebrated types from such great pre-grouping companies as the Caledonian and North British vied with my determination to see rarities from the former LMS/LNER, for by this time many of my spotting needs in the more modern classes were inevitably Scottish.

And then in 1955, at the age of 15, having visited most cities of England and Wales, my father agreed to my undertaking a cycle tour of Scotland with my closest spotting friend Brian Stafford. Actually the tour was a combination of train and cycle. Predictably it was our greatest adventure to date and one which did much to broaden our outlook and develop our sense of self-reliance.

Our tour was to include the Highlands and Speyside for *Ben Alder* and the 'D40's but we began by travelling the West Coast main line to Carlisle and cycling thence to Glasgow, covering the principal sheds en route.

Although the memory inevitably fails in detail the general recollections remain crystal clear. In particular I remember how central Scotland's vast industrial activity combined with a dense and superbly utilised railway network. This reached a peak in Glasgow where the endless locomotive sheds, marshalling yards and sidings were not only a revelation to young eyes but an invaluable experience of the railways' role in society.

We cycled through the infamous slums of the Gorbals, visited the vast locomotive-building suburb of Springburn — though many years were to pass before I really understood the greatness of that suburb — and saw Clydeside in better times. At one time 80 per cent of the ships on the world's oceans were built on the Clyde! We journeyed through the industrial heartland between Glasgow and Edinburgh; mile upon mile of manufacturing intrigue all irretrievably laced together by that great steam communication system.

Many of the small engine sheds were tucked away in heavily industrialised areas but we located them easily by the indefatigable *Locomotive Shed Directory* which gave detailed descriptions of how to find every depot from the nearest station, complete with bus numbers and/or walking times etc. Invaluable as this was, we lacked its natural complement 'The Bunker's Directory', to provide details of the best unofficial way in. A fortune would have awaited its publisher!

Having continued northwards via Perth we reached Aberdeen and were thus poised for our journey through Speyside. We proceeded via Elgin and Keith to Inverurie where the Great North of Scotland Railway had their workshops; the beautiful 'D40's abounding at every location. By this time *Ben Alder* was the last survivor of her class and was dumped at Boat of Garten — the most westerly point of the former GNS system. *Ben Alder* had exactly the atmosphere of mystique that I had imagined years earlier; we saw her on shed with three 'D40's and some of my earliest photographs were of this exotic combination. Other inside-cylinder 4-4-0s in the area at that time were North British 'Glens', Scottish 'Directors' and Caledonian 'Dunalastairs'.

Trips to Scotland followed in 1956/7 and in spite of everything I saw, only a little photography was done and the tragic loss of those negatives in the intervening years means that this Scottish chapter consists principally of work done on the 1965 trip which was largely inspired by the last of the LNER 'A4' Class 'Pacifics'

being transferred from the East Coast main line and put to work on the Glasgow Queen Street/Aberdeen run.

At this crucial time the last LNER 'A2' and 'A3' Class 'Pacifics' could be found in nearby Edinburgh along with a handful of North British designs which miraculously survived further north on the Fifeshire Coalfield. But sadly, by the time my camera was really brought into play, all representatives of the Caledonian Railway, Glasgow and South Western, Highland and Great North of Scotland had passed into history.

Below *One of my rarest prints featuring a survivor from the Caledonian Railway taken on the Scottish tour in 1956. At that time, there were some 6,000 Caledonian engines still in BR service—about half the quantity acquired by the LMS at the grouping of 1923. They were mainly inside-cylinder 0-6-0s and 0-6-0Ts and all had disappeared by November 1963. This engine, No 56262, is one of McIntosh's standard shunting tanks built between 1905 and 1922. She is seen at Inverness.*

Above right *A scene on the busy ash pits at Inverness depot in 1957 with former Caledonian Railway 'Dunalastair' 4-4-0 No 54484— one of a later batch of 48 engines built between 1916 and 1922. The dumper wagon is one of several used to convey ashes taken from the pits.*

Below right *Most pre-grouping engines remained on their home territory, but one notable exception was the transfer during LNER days of two of these ex-Great Eastern 'F4' Class branch and suburban engines from their London (Liverpool Street) and East Anglian haunts to Kittybrewster shed, Aberdeen. They were sub-shedded out to Fraserburg for working the lonely St Combs branch and carried special cowcatchers for the purpose! No 67157 was the last survivor and is seen here at Kittybrewster works in 1956 having just been withdrawn from service.*

Below *One cannot speak too highly of Gresley's three-cylinder 'V2' 2-6-2s. Excellent on express passenger or fast freight, they were referred to as the 'engines that won the war' owing to their prestigious feats of haulage under austere wartime conditions. In many respects they were the equal of the 'A3 Pacifics'. The 'V2's were originally introduced in 1936 for the celebrated 'Green Arrow' fast goods service from Kings Cross to Scotland. Chaotic transport policies over recent years have prevented the railways from providing such excellent services.*

Right *The superb thoroughbred lines of a 'V2' complete with 6ft 2in diameter driving wheels. The class consisted of 184 engines which remained intact until withdrawal commenced in 1962.*

Above *The two mixed-traffic stalwarts of the LNER were the 'V2' 2-6-2s (right) and 'B1' 4-6-0s (left). Together, the two classes totalled almost 600 engines. When the 'B1's were introduced in 1942, it was intended that they should replace a plethora of ageing 4-6-0s of various kinds totalling 387 engines of 21 different types!*

Right *Happy times at St Margaret's shed, Edinburgh, as I proudly pose alongside Gresley 'A3 Pacific' No 60041, Salmon Trout. During the great train spotting years, this engine had been allocated to Edinburgh Haymarket depot and was a notoriously 'hard' 'A3' for spotters in the Midlands.*

Below *Gresley's 'A3's were every bit as graceful as the famous racehorses they were named after and their scintillating performances over the East Coast main line truly lived up to the association. In later years the 'A3's acquired double chimneys and German-style windshields, modifications which—though imposing—greatly marred their original beauty.*

Below *When we toured Scotland in 1965, No 60041, Salmon Trout, was one of the last two surviving Gresley 'A3's; she was withdrawn in December of that year, but stood dumped for some months and was actually the last 'A3' to be broken up. She finally disappeared into Arnott Young's Carmyle scrapyard in November 1966.*

Above right *Withdrawn engines were sometimes retained as 'stationary boilers' to supply steam to workshops or other installations. Often, the locomotive would be complete and many engines survived long after their natural time span on such duties. Here we see former North British 'J36' Class 0-6-0 No 65234—note the steam supply pipe leading from the dome.*

Below right *Another 'J36' consigned to supplying steam was No 65327. I always found these stationary boilers eerie; a complete engine which had not turned a wheel in years would stand panting, hissing and smoking in a lonely siding surrounded by pipes and debris. Often such engines were totally isolated—the track having been lifted behind them!*

SCRAP BRAKE
BLOCKS
ONLY

SCRAP STE
ONLY

Above *A fine example of a latter-day suburban engine in the form of an ex-LMS Fairburn 2-6-4T, almost 300 of which were built between 1945/51. After nationalisation, these sprightly engines could be seen from Kent to Central Scotland and they formed the basis for the BR standard 2-6-4Ts.*

Below *Sunlight reacts vividly with the smoke and steam rising within the magical portals of St Margaret's shed Edinburgh. A simple everyday scene of the steam age. The engine, No 80054, is one of BR's standard 2-6-4Ts introduced for fast suburban work in 1951.*

Above *The Gresley 'A4's shared some duties on the Glasgow—Aberdeen run with the BR 'Standard 5's. Here No 73153—one of 30 fitted with Caprotti valve gear—prepares to depart from Perth in July 1965. She had been built exactly eight years earlier as the penultimate steam locomotive to emerge from Derby works. Withdrawn in December 1966, she was broken up by Shipbreaking Industries of Faslane in April 1967 after a ridiculously short working life of only nine years.*

Below *Hornets Beauty, a former LNER 'A2' Class 'Pacific', at Polmadie shed Glasgow in July 1965—the month she was withdrawn from service. She was broken up at the Motherwell Machinery and Scrap Company in Wishaw exactly one year later.*

Left *There was a great mystique about the LNER 'A2 Pacifics'. The class totalled 40 engines decked with names guaranteed to inspire anyone's imagination. The 'A2's were distributed between eight depots ranging from Aberdeen to Peterborough but as the type was often used on shorter haul expresses—or even fast freight— the North Eastern and Scottish ones seldom worked to London and despite many long vigils beside the East Coast main line I never did see No 60539,* Bronzino, *a Newcastle (Heaton) engine. Another traditionally 'hard' 'A2' of those childhood years was No 60530,* Sayajirao, *seen here at Dundee Tay Bridge shed with the more famous* Blue Peter *during the summer of 1965. They were the last survivors of the class;* Blue Peter *was to be preserved, whilst* Sayajirao *was broken up at Motherwell Machinery and Scrap Company in March 1967.*

Below left *I was always fond of these little Ivatt 2-6-0s despite their being responsible for ousting many old and exotic pre-grouping types. The class was introduced by the LMS in 1946 and No 46462 was one of a batch distinguished by their tall thin chimneys as compared with the standard ones which were shorter and broader (see page 28). She was a regular Edinburgh (St Margaret's) engine all her working life.*

Below *The North British Railway relied on inside-cylinder 0-6-0s for heavy freight and mineral hauling—particularly Classes 'J36/7'. After the grouping, Gresley produced a larger 0-6-0 in the form of 35 of these imposing 'J38's; all were built at Darlington in 1925. Excellent mineral haulers, the 'J38's appeared extensively on the Fifeshire coalfield and survived there until 1965 when I caught No 65932 beneath the coaling plant at Thornton Junction.*

Left *To cope with increasingly heavier coal trains, the North British Railway's ubiquitous 'J36's (page 26) were augmented with the more powerful 'J37's, 104 of which were put into operation between 1914 and 1921. Here is No 64570 reposing in the depot yard at Thornton Junction.*

Above *A 'J37' Class 0-6-0 takes a breather between duties at Thorton on the Fifeshire coalfield. Several dozen of these pre-grouping veterans survived on mineral hauls in Scotland until as late as 1965.*

Below *Having spent years enjoying the LNER 'Pacifics' on the East Coast main line, dieselisation came as a blow. Then came the news that some 'A4's were to be transferred to Scotland for working the Glasgow (Queen Street)—Aberdeen service. The 'A4's were the highspot of our Scottish tour in 1965 and here No 60019, Bittern, pulls out of Queen Street with an express for the Granite City. A Gateshead engine for many years, Bittern was transferred to Aberdeen (Ferryhill) in October 1963.*

Left *With a deafening roar of steam from her cylinder cocks, 'A4 Pacific' No 60019,*
Bittern, returns to Glasgow's Eastfield shed having brought in an express from Aberdeen.

Below left *We called her, 'the commonest streak on the main'. From my first visits*
to the East Coast main line at Grantham, I remember seeing 'A4' 60026, Miles
Beevor. She was a Peterborough (New England) engine and truly earned the
juvenile nickname 'Crate'—often appearing twice in one day! Little did we know
how much we would revere her 15 years later when steam traction was rapidly
becoming extinct. The small boys are certainly delighted and seem intent on 'cabbing
her' during a brief stop at Perth with an Aberdeen—Glasgow express.

Below *'A4 Pacific' No 60026, Miles Beevor, glistening in the late afternoon sunlight*
at Glasgow Queen Street station one summer evening in 1965. She had been
transferred to Scotland from the southern reaches of the East Coast main line in
October 1963. The engine was withdrawn six months after this picture was taken
and stood in store at Aberdeen Ferryhill for a year. In 1967 she spent a further year
in Crewe works providing spares for sister engine No 60010, Dominion of
Canada—scheduled for preservation. At the end of 1967 the remains of Miles
Beevor were taken to Hughes Bolckows of North Blyth and broken up in
January 1968.

Following page top *This delightful 'maid of all work' belonged to the 'J36' Class*
comprised of 168 former North British Railway freight engines built between 1888
and 1900. All were built at the company's Cowlairs works in Glasgow with the
exception of 15 from Neilson in 1891 and 15 from Sharp Stewart the following
year. During World War 1, 25 were shipped to Europe and upon their return were
given names to commemorate famous people and events from the war. This was
one of the few occasions in British locomotive history when names were given to
the humble inside-cylinder 0-6-0 tender engine.

The Ben pictures

Below Left Ben Alder, *the engine which inspired so many boyhood dreams, finally run to earth at Boat-of-Garten—a sub-shed to Aviemore. This was one of my first photographs but one which is particularly historical as* Ben Alder *was the last surviving member of the Highland Railway's celebrated 'Small Ben' Class comprised of 20 engines built during the late Victorian and early Edwardian period.*

Right *The lovely Victorian face of Ben Alder. After withdrawal in 1953, the Ben stood at Loch Gorm works before being moved to Boat-of-Garten pending possible preservation. She was later stored at Grangemouth shed in Glasgow where I found her on the 1965 tour. I was never convinced that she would be preserved and, sure enough, she was tragically broken up shortly after this picture was made.*

Chapter 2

Carlisle

Carlisle has never failed to inspire lovers of railways as the meeting point of seven different railway companies all of which had distinctive designs of locomotive and rolling stock along with different liveries. The list is imposing and it includes some of the finest pre-grouping railways: London and North Western, Midland, Maryport and Carlisle, North Eastern, Caledonian, Glasgow and South Western and North British.

By 1955 however, though still a magnificent railway centre boasting four sheds, little of the former grandiloquence was evident. In fact the former Midland Railway Shed at Durran Hill was almost abandoned and

it was here that I experienced a genuine derelict in an exceedingly rusty and overgrown state. The engine was No 52418, an old Lancashire and Yorkshire railway goods engine and its eerie atmosphere created a profound emotional reaction as it was unusual to see such derelicts at that time. Normally, engines were either stored in greased condition with sacks over their chimneys, or cut up within months of withdrawal. Experiencing this derelict was the beginning of a photographic ideal which I was to express on a world-wide canvas over later years. Also at Durran Hill had been four Tilbury 4-4-2Ts which we had seen earlier

These Ivatt 2-6-0s consisted of 128 engines which were a spotter's nightmare. The class was distributed between some 40 different sheds ranging from Aberdeen, in Scotland, to Northumberland, Wales, East Anglia and the Midlands through to the South-West of England. As branch and trip engines, many were sub-shedded out to places remote from their home depots which made some engines extremely elusive.

that year at Derby works for scrapping and they had come covered in rust, soil and weeds.

Durran Hill's duties were transferred to the former London and North Western Depot at Upperby one of the West Coast main line's great depots. Upperby was exciting to visit as it always produced rarities from Central Division and Scottish sheds which had no regular workings to the south.

However, at the former North British shed at Carlisle Canal we experienced types hitherto unseen; particularly the 'J36' Class 0-6-0 freight engines introduced in 1888. Some of these engines worked overseas during World War 1 and were named after famous military leaders. The names were crudely painted on the central splasher and one of Canal's engines was No 65216 *Byng* under which a local wag had etched 'Crosby' and I clearly remember my disdainful reaction to so dignified an engine being affronted in this flippant way! Also at Canal I saw my first North British 'N15' Class 0-6-2Ts.

And thus to mighty Kingmoor to be enraptured by delights ranging from rare 'Jubilees' to old Caledonian classes of both tank and tender variety which were at the southern limit of their range. It was on large tours such

as this that one really appreciated the way in which most pre-grouping classes remained on their original territories over 30 years later. Apart from creating fascinating contrasts from shed to shed and throughout different parts of the country, this situation enabled students of railway history to appreciate the structure of the former private companies. This memorable visit was the first of many to Carlisle and the last, in 1965, produced the pictures in this chapter. By then, only Kingmoor and Upperby survived and today only Kingmoor.

Carlisle's great railway past renders it an inevitable focal point in the constant running down of Britain's railway industry. The battle to prevent the closing of the former Midland Railway's main line from Settle to Carlisle is well known enough but the wholesale ripping up of the great marshalling yard at Kingmoor — a mere 20 years old — is a pernicious act which has gone largely unnoticed. Closure of this yard typifies the defeatism rife within BR.

And so, these few words upon Carlisle are written with nostalgia and anger; emotions which possibly lend a little extra meaning to the assembled pictures.

A distressing scene at Kingmoor shed depicting three withdrawn 'Jubilee's complete with bags over their chimneys. They are right—left; 45588 Kashmir, 45629 Straits Settlements, and 45742 Connaught. Kashmir had been a regular performer on the West Coast main line; Straits Settlements a favourite at my childhood bridge from 1949, whilst Connaught was once famous for her double-chimney and fine performances on the London (Euston)—Birmingham—Wolverhampton expresses.
Within weeks of this picture being made, the three were towed to Motherwell Machinery and Scrap Company Wishaw and broken up in August 1965.

Below *A mixed freight leaves Carlisle for Glasgow behind BR 'Standard 5' No 73102. Built in Doncaster in 1955, No 73102 was a Glasgow (Corkerhill) engine for her short working life of eleven years. Withdrawn from Corkerhill in December 1966, she was broken up by Campbells of Airdrie the following April.*

Right *BR 'Pacific' No 70029,* Shooting Star, *at Upperby shed. Originally one of the Western Region's Cardiff (Canton) 'Britannias',* Shooting Star *once worked such crack London expresses as 'The Red Dragon' and 'The Capital's United'. The 'Britannias' were built between 1951–54 and performed distinguished service on all regions until their brilliant career was cut short by dieselisation—which forced them to be relegated to secondary work.*

Below right *The imposing lines of the BR 'Clan Pacifics' are evident in this study of 72007,* Clan Mackintosh, *at Kingmoor. The 'Clans' were a lighter version of the 'Britannias' and ten were built in 1952 for the Scottish Region. Further deliveries were cut short by dieselisation. Withdrawal commenced in 1962, but No 72007 remained active until November 1965 and finally disappeared into Campbell's Airdrie yard in March 1966.*

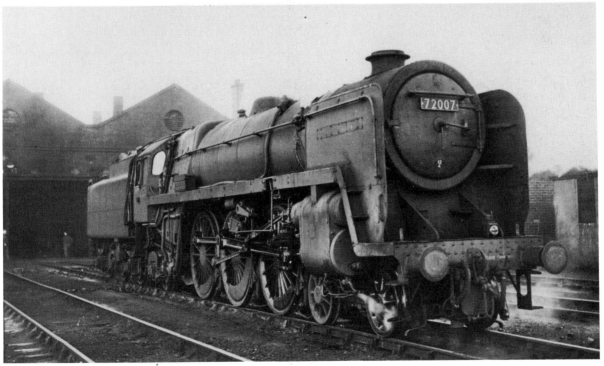

A distinguished survivor in the repair bay at Kingmoor in the form of 'Scot' Class 4-6-0 No 46115, Scots Guardsman. From their inception in 1927, the 71 members of this class did premium service over the West Coast main line especially after their rebuilding with taper boilers. They were named after famous regiments— invariably with the regimental badge over their nameplates. No 46115 was the last survivor and ended her days on local freight trips along with a turn from Carlisle to Glasgow Moss End.

The slogging haul over Beattock was the scourge of enginemen on the Carlisle—
Glasgow run. The bank lies some 40 miles north of Carlisle on the former
Caledonian Railway's main line. Banking engines were often employed and during
the mid-60s this duty was entrusted to a stud of LMS 'Fairburn' 2-6-4Ts.

Above *'Britannia' No 70005,* John Milton, *at Citadel Station.* John Milton *was originally one of the Great Eastern section 'Britannias' whose work on the two-hour London (Liverpool Street)—Norwich expresses—complete with sprints of 100 mph—has gone down in railway history. By 1965, No 70005 was a Kingmoor engine; she remained in service for two years after this picture was made before being scrapped by Campbells of Airdrie in January 1968.*

Below *The unmistakable wall of Citadel Station makes an imposing backdrop to ex-LMS Ivatt 2-6-2T No 41217 built at Crewe in 1948.*

Chapter 3

North Notts/South Yorkshire coalfield

This vast coalfield gave rise to the coal hauls to London as typified by the famous Toton–Brents. If one also considers the extensive coal drags over the Great Northern main line and from the turn of the century, the heavy traffic south along the Great Central, it would be no exaggeration to say that several billion tons of coal have been conveyed to the south of England from this field by steam power.

Over the years a fascinating variety of motive power evolved to handle this traffic, beginning with the notorious double-heading of inside-cylinder 0-6-0s over the Midland main line — a fascinating practice which continued until the advent of the LMS 'Garratts' in 1927 and the Stanier '8F' Class 2-8-0s seven years later. In their final years the Toton–Brents were handled by BR '9F' Class 2-10-0s both with and without Franco-Crosti boilers.

On the Midland main line, coal trains headed south and the iron ore from Northamptonshire headed north to the founderies and ironworks built on the Nottinghamshire/Yorkshire coalfield. The hub of this vast industrial complex was Sheffield — 'hell with the lid off' as it was known at the height of its sulphurous past — and coal and iron ore formed a considerable percentage of the traffic on the Midland main line.

I recall the many visits made by train to Toton, Westhouses, Hasland and Staveley. The expresses from St Pancras to Leeds would crawl through the Erewash Valley carefully observing the restrictions caused by mining subsidence. Heading northwards from Toton, we soon passed the innovative steel works of Stanton & Staveley before entering a wonderland of heavy industry; mile upon mile of collieries, foundries and

factories, amid a vast Lowry-like terrain covered in sidings and branch lines, with plumes of steam rising into the air at all points of the compass from a myriad shunting engines.

Sheffield, bedecked with rolling mills and steel based fabrications, was once the cutlery manufacturer to the world. How nobly this great city was commemorated by the Great Central line's Sheffield to London Dining Car Express, 'The Master Cutler'. How magnificent the Gresley 'A3' would look with its fine headboard and long rake of varnished teak coaches. Today a train still runs with that title but from the old rival station of St Pancras — it has no restaurant car and BR no longer has the energy or imagination to don a headboard.

To the north of Sheffield is Wakefield, once a vast clearing house for the South Yorkshire Coalfield and a depot famous for its allocation of 60 'WD Austerity' 2-8-0s.

Commensurate with the run down of steam, came a marked waning of activity; many collieries and ironworks closed, domestic coal burning declined whilst the steam locomotive itself — one of coal's biggest customers — had largely disappeared from the south, thus eliminating the need to convey coal to the huge number of southern depots situated in areas devoid of indigenous supply. Although closure of the Great Central was imminent in 1965, a handful of the classic LNER mineral engines survived and it was these that prompted my trips to Staveley to make these pictures on two cold winter days of that year. Featured are Robinson's un-rebuilt '04' Class, Thompson's rebuilt '01' Class along with the ubiquitous 'WD' Class 2-8-0.

Seeing the Thompson '01/04's for the last time, I

recalled their intensive south-bound coal hauls over the Great Central. Annesley shed, north of Nottingham, had 50 allocated for the purpose and to this day I can hear the girders of the huge bridge which carried the Great Central over the west-coast main line at Rugby rumbling deafeningly as a heavy coal train raced over behind one of the 'Annesley Rods' as the type was commonly known.

Amazingly, Staveley Ironworks retained five ancient Midland Railway '1F' Class 0-6-0Ts for shunting duties along with a Deeley 0-4-0T. Apart from their suitability for operating in restricted areas of the works, it is believed that their retention was the result of a hundred year agreement between the Midland Railway and the ironworks in 1866; all were withdrawn when the agreement expired. These historic '1F's were the immediate forerunner of the famous 'Jinty' the standard shunting type on the LMS.

Below *My visits to the Nottinghamshire coalfield produced the last survivors of the celebrated Robinson '04' Class. Introduced in 1911, these 2-8-0s were the Great Central's principal freight type and following the outbreak of World War 1, the design was adopted by the Railway Operating Division (ROD) for service overseas.*

Right *One summer day in 1954, I cycled to Heanor Colliery on a most exciting mission; I was going to see* Cecil Raikes—*an engine which fascinated me so much that I actually took my camera with me!* Cecil Raikes *was one of nine massive 0-6-4Ts built by Beyer Peacock in 1885 for the Mersey Railway's underground line between Liverpool and Birkenhead. When built, they were the most powerful locomotives in Britain. The Mersey Railway was electrified in 1904 and the giants were sold off; three to the Alexandra Docks Railway; four to Australia and two to Shipley Collieries, Notts.* Cecil Raikes *was the last survivor and here is one of the pictures I made on that unforgettable day. Notice that she still bears the huge condensing pipes although they would not have been used for 50 years.*

Below right *One of the last survivors of 240 former Midland Railway shunting engines built between 1878 and 1898. The forerunner of the more familiar 'Jinty' (page 120) these '1F' tanks—as we called them—survived on shunting duties at Stavely Ironworks until 1965 when this one, No 41712, was despatched to Cohen's at Kettering for scrap. During her 85-year working life, she had been rebuilt with a Belpair boiler, but retained her original chimney and half-section cab.*

Left *The 'J50' Class 0-6-0T shunter was one of Nigel Gresley's lesser known designs and appeared during his years as Chief Mechanical Engineer of the Great Northern Railway. Introduced in 1913, building continued until 1937, when the class totalled 102 engines. The 'J50's were ubiquitous throughout the former GN system until the early 1960s when withdrawals were rapid. By 1963, only seven survived, all in departmental service and here—as Doncaster works' No 12—is former No 68917, which I remember as being a Hornsey engine.*

Below left *This little known Midland Railway class consisted of ten engines designed by Deeley and built between 1907 and 1922 for shunting in docks, brewery yards and other industrial establishments where tight curves could be found. No 41528—the oldest member of the class—survived at Stavely Iron Works until the mid 1960s when this picture was made.*

Below *As a result of the Great Central's '04's being utilised for military service overseas during World War 1, the class reached a total of 521 engines. Most of these were eventually absorbed into the LNER network, but some passed to the GWR and even the L & NWR! The GW examples survived until 1958, but the 50 allocated to the L & NW, had, amazingly, disappeared by 1933. The '04's saw further overseas service during World War 2, with certain engines being 'called up' for a second time.*

Left *A scene in the works' yard at Doncaster depicting three Gresley 'J50' 0-6-0Ts withdrawn from line service but retained for a further lease of active life as departmental engines for operations within the works confines.*

Below left *One bitterly cold morning with an icy wind blowing, I did a stint of lineside photography at Whetstone on the former GC main line. It was sunny and the low temperature ensured splendid exhaust effects. The best picture made that morning was of a southbound coal haul from Nottinghamshire headed by BR '9F' No 92092.*

Below *Although many Great Central '04's remained in their original condition, others were subjected to various modifications and rebuilds which resulted in seven different varieties, adding considerable technical, historical and aesthetical interest to the class. Here is No 63691 rebuilt with a 'B1' round-topped boiler and chimney but retaining her original cylinders and running plate.*

A smoky scene in Doncaster shed depicting 'War Department' 2-8-0 No 90484—
one of twelve once allocated there. In the background is a former Great Central '04'
rebuilt with round-topped boiler.

Contrasting giants at Doncaster Motive Power Depot featuring (left) a Robinson '04'
Class rebuilt with a 'B1' type round-topped boiler and a pair of 'WD Austerity' 2-8-0s.
These two classes alone originally totalled 1,254 engines—an indication of the vast
freight traffic once handled by our railways.

Chapter 4

Leicester and the Midlands

I was lucky to be born in Leicestershire, since apart from being well blessed with railways, the county lies in the centre of the country and is thus strategically placed for travel. It was in Leicestershire that I first experienced the spell of the railway one summer afternoon in 1949, when after school in Oadby, I cycled with a friend over the fields to the Victorian bridge at Newton Harcourt, a tiny village located on the Midland main line seven miles south of Leicester. I was subsequently to spend much of the next two years at that enchanted bridge witnessing the joys of a busy steam main line with virtually every type of train imaginable.

From that bridge at Newton Harcourt I began my travels across the country; initially by making short trips to Rugby on the West-Coast main line; Grantham on the East-Coast main line or Derby shed and works. But soon I was to travel throughout Britian and subsequently, of course, the whole world.

Today, 35 years later, living in the cottage which perchance overlooks the very bridge, is a constant reminder of my roots. Sitting by that bridge as a child I never dreamt that from the cottage but one field away, I would mount expeditions to scour the earth in search of the sights which were then being witnessed daily.

During those formative years, innumerable forays were made to the sheds in Leicester; the Midland, Great Central and Great Northern. The Midland was the favourite with an allocation of some 85 engines, ranging from superannuated ex-Midland Railway inside-cylinder 0-6-0s dating back to 1875, through to the latest BR 'Standard 5' 4-6-0s newly built at Derby works. Above the depot yard was a snicket which provided a short cut between two main thorough-fares. Known as

the Bird Cage, this path commanded a breathtaking view of both the shed yard and the main line and was the premier spotting location for Leicester. On summer Saturdays during school holidays up to 150 'kids' would be congregated at any one time.

Amongst the wide diversity of trains were certain off-beat diagrams which had the potential of bringing unusual or rare engines from distant sheds and at the Bird Cage we eagerly awaited the 'Kingmoor' ('4F's, 'Black 5's, 'Crabs' or 'Jubilees' from Carlisle) the 'Edge Hill' or the 'Ancoats' — which sometimes brought a rare Central Division 'Crab'. My favourite however was the '5A' which came as far as South Wigston from Nuneaton. We never found out the other parts of this diagram but around 4.30 pm each weekday a short pick-up freight arrived from Nuneaton, headed by anything from an un-rebuilt 'Patriot' to a Stanier 2-6-0. Frequently it was a 'Patriot' and we regularly cycled over from Leicester to see it before it returned around 6.00 pm. The bridge carrying the Midland main line commands an excellent view over South Wigston and I can still see in my mind's eye the familiar parallel boiler, flat windshields and capouchon chimney of an un-rebuilt 'Patriot' protruding above the wagons of the distant yards. What delights from faraway places this exotic working brought to Wigston; No 45502 *Royal Naval Division*, No 45503 *Royal Leicestershire Regiment*, No 45504 *Royal Signals*, No 45507 *Royal Tank Corps*, No 45533 *Lord Rathmore*, No 45543 *Home Guard*, or one of the unnamed 'Patriots' such as No 45550. But the '5A' was quite unpredictable and on one occasion a Perth 'Black 5' turned up complete with snow plough!

The Great Central was a thriving line in those

boyhood years; it provided an alternative and competing route between London and Leicester, Nottingham, Sheffield and Manchester. The principal expresses were worked by Gresley 'A3's including No 60103 *Flying Scotsman* which was a Leicester engine for some years. Certain interesting workings often took us down to the Great Central especially the daily 'Western' which brought a Great Western 'Hall' up from Banbury each afternoon and the 'York Goods', a fully-fitted freight which was often hauled by one of the tantalisingly beautiful Raven North Eastern 'B16' Class three-cylinder 4-6-0s. The Central was a magnificent railway in every respect and I remember my grandfather telling me he watched the huge girder bridges being craned into place because, Leicester being well developed by the turn of the century, much of the Central — the last main line into London — had to be constructed on high abutments.

My memories of the Great Northern depot are vague; it offered nothing in the way of big engines but at the age of eleven my instincts for locomotives of distinction were beginning to develop and I used to savour seeing the ex-GN 'J6's and even older 'J5', 0-6-0s which worked in along with 'Ragtimer K2's from Colwick.

Leicester's railway history however began almost three-quarters of a century before the Great Central, when the Stephensons engineered the Leicester and Swannington Railway to bring coal to the city from the Leicestershire coalfield. Opened in 1832 the line incorporated the second railway tunnel in the world and featured the first inside-cylinder 0-6-0s, the first of a long dynasty which was to epitomise the definitive Britsh steam locomotive. A further distinction on the Leicester and Swannington Railway was the first locomotive whistle — or steam trumpet — allegedly applied after one of the locomotives had careered through a cartful of farm produce on a road crossing. Rather better known perhaps is the first excursion train run by Thomas Cook between Leicester and Loughborough in 1841.

For many years before the Swannington line closed, the mile-long tunnel was declared the oldest working railway tunnel in the world and until the early 1960s the line was worked by the last Midland Railway '2F' Class 0-6-0s from Coalville depot, as these were the only BR engines strong enough yet small enough to traverse the incredibly narrow bore.

All too soon there were only memories; the Great Central and Great Northern had gone, whilst steam workings over the Midland main line had almost disappeared. One of the last turns was the 2.35 pm freight from Leicester to Wellingborough.

The mid 1960s were a time when Cohens scrapyard at Kettering came to prevalence as one of the main railway breakers in Britain and the sight of condemned engines standing in sidings on the south-western side of Kettering station became a familiar spectacle. I became doubly sad when the inmates were old Leicester line favourites but many other types appeared, often from afar, and amongst the LMS '1F's, '4F's and 'Black 5's were LNER 'B1's; GW 'Castles' and Southern 'W's, 'Schools' and 'Pacifics'.

I was thrilled to see the last of the London & North Western Railway's 'Gobblers' and my favourite was No 46604. Whenever I went to Rugby as a small boy she was there. She operated the two-coach Warwick and Leamington trains. The Gobblers worked branch and cross-country connecting lines throughout the former L&NWR network; 160 were built during the 1890s and No 46604 was one of the last two survivors. She was withdrawn in 1955.

Left *Leicester '2F' No 58298 was a regular performer on the smelly bone train which conveyed several wagons of bones from Leicester Cattle Market to the glue factory at East Langton near Market Harborough. The 'Smelly Bone' used to come through Newton at lunchtime; on hot days the stench was terrible and I used to refrain from eating my sandwiches until after the 'Bone' had passed. The '2F' would return running tender-first with empty wagons around 4.00 pm.*

Below *By 1950, Leicester's former Midland Railway '2F' 0-6-0s were primarily relegated to shunting, either at the north yards, Knighton or Wigston and were a typical example of 19th century goods engines being down-graded to humble tasks in the twilight of their lives. No 58305 was not one of Leicester's traditional '2F's as she came to us from Bedford during the mid-1950s.*

Right *For as long as I can remember stored engines had sacks placed over their chimneys; a simple act, but one which added an aura of mystique. I have never known exactly why engines were bagged in this way, neither do I know how long the practice had been in operation before I first experienced it in 1950. It is a fascinating phenomenon and one that I have never witnessed since in my travels overseas.*

Left *A view from the Bird Cage at Leicester featuring BR 'Standard' 2-6-0 No 78027, with 'Jubilee' No 45573, Newfoundland. This 1964 scene was made when steam and diesel were equally divided at this once busy depot.*

Below left *Over the last ten years of British steam, vast numbers of locomotives were put out to tender for breaking up by private firms. One of the innumerable yards which purchased engines was Cohen's at Cransley situated on the former Midland Railway's ironstone branch from Kettering to Loddington. This brace of LMS '4F' 0-6-0s arrived coupled together and stand at the approach to the breaking up bay.*

Right *I am often asked, 'what is your favourite class of engine?' Without doubt it is the 'Jubilee'. I grew up with them on the Midland main line and apart from being extremely beautiful engines to behold, their names commemorated aspects of Britain's great historical past including all territories of the Empire. I learnt more about British history from the 'Jubilees' than I ever did in the classroom.*

Below *At lunchtime we went down to the Midland sheds; there was a 'Jubilee' in the yard, No 45573, Newfoundland. She had been a Leeds (Holbeck) engine for as long as I could remember and had been a regular engine through Newton Harcourt with such trains as the 'Thames Clyde Express'. But it was now 1964 and as 'Jubilees' were becoming rare at Leicester, I asked for my picture to be taken with an engine I had known so well for 15 years.*

Newfoundland was withdrawn the following year and broken up by Cashmores at Great Bridge in January 1966.

Below *We always associated Burton shed with its 'Crabs', some of which had Lentz rotary cam poppet valves. I used to see these 'Crabs' at Newton on the London beer trains; there were two workings each evening and I always tried to see them before leaving the line, but invariably it would be dark before the second one came. Years later, after the 'Crabs' had gone, a visit to Burton shed produced evidence of a flourishing beer industry, but all that survived on shed was '8F' No 48728—a Leicester engine since the 1940s.*

Right *How one mourns for the days of healthy goods yards and a railway which handled the nation's freight. The Stanier '8F' 2-8-0s of which there were almost 700—were but one of numerous freight classes. Here, No 48637—one of Toton's 50 '8Fs'—prepares a train at Leicester North sidings.*

Below right *The powerful lines of BR's ultimate freight hauler—the 'Standard 9F' 2-10-0s—are evidenced in this study of No 92006 on the ash pit at Leicester Midland shed. The huge concrete coaling plant, which could be seen from many parts of the city, is in the background.*

Below *A rare bird to the Midlands arrived at Kettering in 1964 in the shape of Southern 'W' Class three-cylinder 2-6-4T No 31914. She was one of 15 engines built in 1930/2 for heavy freight tripping in the London area. No 31914 had come for scrap at Cohen's Cransley yard and is seen here in position on the breaking up bay.*

Right *Surrounded by pieces of scrapped engines, former LMS 'Jinty' No 47501 awaits breaking up at Cohen's yard at Cransley. This 'Jinty' was allocated to Devon's Road (Bow) for many years and her superb condition indicates that her withdrawal was merely a formality amid the desire to be rid of steam.*

Below right *A stranger to Nottingham shed in 1964 was Great Western 'Castle' Class 4-6-0 No 7029, Clun Castle on one of her frequent railtours. As Britain's surviving steam locomotives became increasingly begrimed and run down, Clun Castle was resplendently maintained for special services and over 1964-65 literally millions of people turned up to the lineside to see this beautiful locomotive. When she worked the last steam train from Paddington to the West Country in November 1965, an endless sea of faces—many tear stained—lined the route.*

Newton Harcourt

Below *After regular steam workings had finished through Newton, a few specials passed until BR finally banned them.* Flying Scotsman *made several appearances to the delight of many villagers at that time and she is seen here in full cry. In all my childhood years, I had never before seen a Gresley 'A3' through Newton.*

Right *As a small boy I stood talking to a group of footplatemen in a mess room at the north end of Rugby Station and one senior driver said; 'the Stanier 'Black 5's could work any train on the system'. 'Do you think so?' I asked. 'Think so, I bloody well know so!' he bellowed. One heard similar sentiments expressed almost daily during the early 1950s and here one of this exalted class dashes through Newton Harcourt with a fitted freight.*

Below right *A 'Castle' on the Midland main line! During one of No 7029 Clun Castle's many railtours, she worked a special from Nottingham to Northampton. She is caught here in full cry at Newton Harcourt as she attacks the long southbound climb from Kilby Bridge to Kibworth.*

Below *Time came when the steam train no longer ran through my childhood village and by 1965 the music of a three-cylinder 'Jubilee' on the climb up from Kilby Bridge would never be heard again. But the bark of the '8F' lingered a little longer, as witness No 48670 storming through Newton Harcourt's snow covered cutting with the 2.35 pm Leicester—Wellingborough freight.*

Below right *. . . this working was one of the last regular steam turns southwards from Leicester and having passed the camera, No 48670 prepares to burst beneath the village bridge amid a glorious pall of smoke and steam which would characteristically obliterate most of the village square.*

Right *The railway abounded in a plethora of fascinating detail on which the eye could rest between trains and just beyond my spotting bridge at Newton Harcourt, was a Midland Railway ¼ mile post and a set of catch points. These points would derail any wagons which broke away from a loose coupled freight on the climb up to Kibworth and so prevent them from running into an on-coming train.*

Leicester and Swannington

Below *As a boy, I was always proud that Leicester was the richest city in Europe and fragments of its once fine manufacturing traditions can be seen as a back drop to this BR 'Standard' 2-6-0 at West Bridge.*

Right *Local merchant, W. Baines & Co, collects coal from the wharf as BR 2-6-0 No 78028 shunts the sidings. A fine example of the proper use of railways at Leicester West Bridge.*

Below right *By 1964, the 132-year-old Leicester and Swannington line from Desford to Leicester had become a railway backwater. But although this scene at West Bridge is lonely, notice the number of lines—at least ten—more than would be found in many important main line sidings on BR today!*

Above *My favourite Leicester '3F' was No 43728—one of 14 of these former Midland Railway freight engines allocated to us in 1950. They were used primarily as shunters particularly at Knighton and Wigston—a sub-shed of Leicester—but they remained in main line service as well with light goods trains and the occasional holiday excursion.*

Above right *On many nights I was lulled to sleep by the sound of these '3F's shunting at Knighton sidings. On damp evenings especially, the barking exhaust beats—and the musical clanking of wagons—floated through my bedroom window in Oadby over two miles away. Here however, '3F' No 43728 is engaged on a special passenger working along the former Leicester and Swannington line and is seen at Glenfield station.*

Right *The former Leicester and Swannington Railway's line included the notoriously narrow Glenfield Tunnel and the only type capable of passing through it was the ex-Midland '2F' with original cut-away cab. Coalville retained three of the type for this service and they survived until 1964. The last one to go was No 58182 of 1876, which at 88, was the oldest locomotive then running on British Railways.*

Great Central

Below *The Great Central, the last main line to London, opened in a blaze of magnificence with excellent services, elegant trains and breathtakingly handsome locomotives. Indeed, its fastest expresses to Sheffield were little slower than BR's Inter-Cities three-quarters of a century later! The destruction of the GC needs little mention here and the last services were operated by run down 'Black 5's displaced by diesels elsewhere.*

Right *Leicester Central Station, a mere 65-years-old and on the point of closure. What a splendid variety of services once emanated from here, both main line and cross country! The diversity of trains brought locomotives from each of the 'Big Four' regions, especially during the peak holiday season. But in 1965, a lonely 'Black 5' settles the ghosts as it shunts forward to the water column before continuing south with a Nottingham—Rugby local.*

Below right *On the night the Great Central closed, we went up to Birstall to see the last train go north behind a 'Black 5'. Instead of staging a massive fight to prevent closure, local supporters merely chalked slogans over the abandoned stations. The passing of the last train and subsequent death of a main line proved too sad for Judy Maddock who, overcome with sadness, turned her back to cry.*

LNW 0-8-0 'Weaser'

Below *The characteristic lines of an ex L&NWR 'Weaser' 0-8-0 at Bescot where the last survivors of this once prolific freight type were allocated. Superb haulers, these 'Super Ds'—as they were known by railwaymen—were exciting to experience; their syncopated beat and the high pitched wheezing of steam from their cylinder glands were recognisable from more than a mile distant.*

Right *The unmistakable front of a London and North Western Railway 'Weaser' 0-8-0. Former L&NW engines didn't carry front number plates even after the 1923 grouping and their bare smokebox doors were immediately diagnostic.*

Below right *We made several trips to Bescot to pay our respects to the last L&NWR 'Weaser' 0-8-0s—a class which once numbered over 500 engines. When I began spotting in 1949, Bescot had 40 of the 350 survivors. Fifteen years later, in the summer of 1964, only three remained and they too were withdrawn before the year was out.*

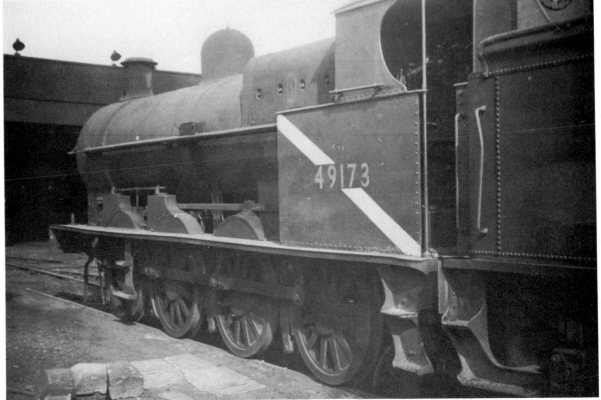

Chapter 5

Great Western days

The last remains of 'God's Wonderful Railway' died along with the rest of British steam and interspersed amongst the classic GW types were BR 'Standards', originally rejected by the region during the early 1950s as being vastly inferior to the indigenous Swindon types — so strong were the traditions of pride and loyalty on this once magnificent railway.

Perhaps these pictures will be particularly interesting in view of the great cult which has developed around the GW's memory over the ensuing 20 years.

In terms of sheer aesthetics, there was little to beat a 'Castle' Class 4-6-0 and my memories of them in resplendent green with gleaming brasswork at Birmingham Snow Hill and Tysley in 1952 are vivid. Actually that first visit to Snow Hill was in an endeavour to see one of the last Bulldogs and although we failed in this, we were lucky to see 'Saint' Class 2932 *Ashton Court*, my only 'Saint' as the survivors disappeared the following year.

Though weaned on the LMS and LNER, I never adopted the partisan attitude of many of my contemporaries and I savoured the GWR with passion. I saw 'Star' Class 4056, *Princess Margaret*, leave Paddington one evening with the five o'clock express to Western-Super-Mare and sister engine 4061, *Glastonbury Abbey,* at Wolverhampton Stafford Road on several occasions. I witnessed Exeter on a busy summer Saturday. How superb those holiday trains were and how puny, trite and self-centred are their present road-choked equivalents. I spotted at Truro for several days during the early 1950s instead of spending time on the beach with my parents, and in between trains sustained myself with genuine Cornish pasties from the station

buffet. And I saw 'Castle' Class 5054, *Earl of Ducie*, travelling at almost 100 mph near Cholsey and evoking poignant memories of 5006 *Treganna Castle*'s 56-minute dash over the 77 miles from Paddington to Swindon with the 'Cheltenham Flyer' in 1932 — an average speed from start to stop of 82 mph. Then to cap it all, I travelled on the last steam train from Paddington on November 27 1965, headed by 7029, *Clun Castle*, the only working survivor. The intention was to run to Swindon, Bristol and Gloucester and upon reaching Swindon on the return journey *Clun Castle* would bow out and two 'EE1750 HP' diesels would dash the train up to Paddington during which an attempt would be made to break the 'Cheltenham Flyer's record of 56 minutes. The tension throughout the packed train on what was an incredibly fast run for the year, was unbelievable; stop-watches pronounced at every mile post our speed and where the 'Cheltenham Flyer' would have been had the two trains been running neck and neck. Almost throughout, we were behind the old Flyer and our final arrival into Paddington was in 61 minutes to the sheer delight of everyone on board. Propaganda for diesels was seldom taken kindly by enthusiasts at that time — unless of course it backfired in the face of its protagonists!

So to my little legacy here; Sunday trips to Wolverhampton Stafford Road and nearby Oxley in 1964 found dead 'Kings' and the 'Castles' reduced to several working survivors, all relegated to secondary duties. Other Sunday trips were to Gloucester; we would start early as we had to be back in Leicester by six o'clock ready to play at the Broken Drum Club in the city centre — our own Sunday night jazz club.

Newport and the South Wales valleys were other districts in which homage was paid, 15 years after I first visited the area to find Taff Vale and Rhymney engines; London and North-Western coal tanks and the incredible array of small tank engines from the old constituent companies which merged into the GWR.

Finally to Barry, one of the most depressing locomotive mausoleums on earth with a large percentage of Great Western types. Barry produced some pictures of types not photographed elsewhere for originally there were 250 locomotives in Woodham's celebrated sidings. Though not appreciated at the time, Barry was to create a renaissance for Great Western glory by allowing time for preservationists to catch up and accumulate funds — all available monies having been dissipated during the scramble to preserve engines before BR broke them up.

One last idyll to the Great Western's memory survived far off its beaten track in the form of *Trojan*, a stalwart 0-6-0 ST from Alders Paper Mills, Tamworth, photographed one bitterly cold Sunday afternoon in the winter of 1966 after all its compatriots had disappeared.

The running-down of the 'Castle' Class was especially tragic. Apart from their elegance and beauty, they were brilliant performers and can be credited with some of the most scintillating running in British railway history. The first 'Castle' appeared in 1923 and the last examples were not completed until 1950—two years after the GW had ceased to exist!

68

Left *From the Edwardian period onwards, faster freights became increasingly important and mixed-traffic engines evolved to fulfil this demand. One important design were the 80 'Granges' introduced in 1936 as a partial rebuild of the earlier '4300' Class 'Mogul' (below). Here is No 6872, Crawley Grange, minus her nameplate, but sporting an '88A' Cardiff (Cathays) shed plate.*

Below *Such was the demand for locomotives around the turn of the century, that 80 mixed-traffic 'Mogul' 2-6-0s were imported from America. The 'Mogul' had been a classic American type for many decades and although the imports were not liked, they hastened the emergence of similar British 'Moguls' for mixed traffic duties. The first railway to respond was the Great Western, when in 1911, Churchward introduced his '4300' Class and so successful were they that building continued until 1932. Here, No 7320 awaits breaking up at Barry in the mid-1960s.*

Right *A milestone in British locomotive history! This imposing freight engine belongs to the famous '2800' Class introduced in 1903. They were the first 2-8-0s to run in Britain and their appearance created a precedent for the type on other railways, although an incredible 32 years were to pass before the LMS adopted a similar design with their Stanier '8F's in 1935 (page 127). The engine illustrated was one of the last batch built between 1938/42; she was withdrawn from Pontypool Road shed in July 1964 and broken up by Buttigiegs of Newport in January 1966.*

Below *A typical GW shed scene depicting one of Churchward's imposing '4200' Class 2-8-0Ts of 1910. Building of these powerful tanks continued for 30 years and they worked the vast South Wales coal traffic from such legendary depots as Newport (Ebbw Jun) Newport (Pill), Cardiff (Canton), Neath and Llanelly.*

Above left *One of the last working survivors of the '2800' Class was No 3864, depicted here at the head of a long coal drag near Newport during Easter 1965. She was withdrawn from Newport shed the following August and broken up in October by Birds of Morriston.*

Left *Work-weary, begrimed and leaking, describes '2800' Class 2-8-0 No 3816. She is seen at Severn Tunnel Junction shed from which she was condemned in August 1965 and broken up by Birds of Long Marston the following December.*

Above *At the height of the industrial revolution, Welsh coal was exported to all corners of the world and an enormous variety of locomotive types emerged to serve the industry. In 1910, the GW increased the power stakes by introducing these large 2-8-0Ts. Known as the '4200's they were, in effect, a heavy tank version of the '2800's and 205 were built. They were almost exclusively involved in the South Wales coal traffic and here No 5209 shunts a colliery interchange against a backdrop of slag heaps.*

Above *It was impossible to travel anywhere on the Great Western system without encountering the ubiquitous 'Pannier Tank'—a form of engine used almost exclusively by the GW for shunting and light tripping. With detail variations, the type was sometimes used on branch line, cross country and suburban work as well. The most numerous class were the '5700's with 863 members built between 1929 and 1949; here, one is caught busily tripping around the yards at Gloucester.*

Above right *Locomotive nameplates were becoming coveted trophies by the early 1960s and although a fine brass plate from a passenger engine could be obtained for as little as £15, it was not difficult to imagine that its value would appreciate. The GW's highly attractive nameplates were particularly vulnerable to theft and most were removed before the engines were actually withdrawn, as in this picture of 'Hall' Class No 6974, Bryngwyn Hall—only the supporting bars being left in place. Today, nameplates from even the most humble types command four figure sums!*

Right *The 0-6-2 tank is normally associated with suburban passenger trains, but following the tradition of the Rhymney and Taff Vale railways, the Great Western made an adaption on the type for short range heavy coal hauling in the South Wales Valleys. Known as the '5600' Class, they were powerful yet sufficiently flexible to negotiate the gradients and sinuous curves on the valley lines. 200 were built between 1924 and 1928. Withdrawal commenced in 1962 and here a condemned example awaits scrapping at Barry.*

Above left *The expansion of cities during the industrial revolution created a pattern of urban living which had become widely established by the late 19th century. Accordingly, the suburban train became a vital facet of railway operation and many forms of tank engines evolved for the purpose. One of the GW's definitive types were these lovely 2-6-2Ts which first appeared in 1903.*

Left Trojan, *and what an amazing career she had. She was built in 1897 by Avonside of Bristol for working in Newport Docks and in 1903 passed to the Alexandra Docks and Railway Company—an organisation responsible for delivering export coal to the loading staithes. The AD was absorbed by the GW in 1922 and* Trojan *became their No 1340. She was withdrawn in 1932, but instead of being scrapped, was purchased by Victoria Colliery of Wellington. In later years,* Trojan *passed to Netherseal Colliery, Burton-on-Trent before being sold to Alder's Paper Mills at Tamworth in 1947. She was still there in 1965 when this picture was made, but shortly afterwards was purchased for preservation.*

Above *The '2800' Class was so important in the evolution of the British steam locomotive that one of them, No 2818, was included in the official list for preservation. She is seen here during her last overhaul prior to display. No 2818 was built in 1905 and withdrawn from Neath depot in October 1963.*

Chapter 6

The Southern, and Britain's last steam main line

My first encounter with a Southern engine was extremely dramatic. I was 10 and an uncle took me to Waterloo Station one night during a family visit to London. I was completely spellbound by the sight of two semi-streamlined 'Merchant Navy Pacifics' No 35010 *Blue Star* and No 35006 *Peninsular and Oriental SN Co*. I was taken into their cabs and talked to the drivers; it was one of the most exciting nights of my life.

My next encounter with a Bullied 'Pacific' was at the age of 11, whilst on holiday at Ilfracombe where the lure of the beach came second place to the station. Bullied's 'Light Pacifics' were almost new at the time and had their original 280 lbs boiler pressures which caused them to blow-off deafeningly. Ilfracombe was a terminus on the branch up from Barnstaple Junction and the little shed always had several 'Pacifics'. My particular favourites were: No 34016 *Bodmin*, No 34024 *Tamar Valley* and No 34027 *Taw Valley*. Ilfracombe was also the destination of the legendary 'Devon Belle' and 'Atlantic Coast Express'.

Forays to the Southern were not over-frequent during my train spotting years but I do remember seeing one of the LBSC 4-4-2s No 32421 *South Foreland* passing over a level crossing while on holiday in the Portsmouth area. This was the only 'Atlantic' I ever saw. Equally memorable was the sight of 'N15 X' Class No 32333, *Remembrance*, along with a breathtaking glimpse of a beautiful 'SEC D' Class 4-4-0 during trips to London sheds during the '50's.

It is paradoxical that the Southern, having been the first railway to adopt extensive electrification as early as the 1920s, should have retained Britain's last steam-hauled expresses. Both in traffic density and locomotive variety, the Southern had always been the least significant of the 'Big Four'; its territory was relatively limited and it embraced few areas of heavy industry. Yet such are the quirks of fate that express passenger trains from London Waterloo to Southampton, Bournemouth and Weymouth remained steam-hauled until July 19 1967, by which time steam had vanished from all but the most menial tasks elsewhere. This remarkable situation was largely due to the modernity of the Bullied 'Pacifics' which had been rebuilt during the late 1950s and proved to be superb performers. Not all were modified however and observers were rewarded by the sight of rebuilt 'Merchant Navys' along with 'Light Pacifics' in both modified and original condition with some BR 'Standard 5' 4-6-0s as well — four types in all.

Inevitably the Southern main line became a mecca for enthusiasts from all over Britain; not least because of its purity as an all-steam railway, for most other steam lines were heavily infiltrated by diesels, which to the vast majority of enthusiasts at that time were thoroughly obnoxious — how feelings have changed during the intervening 20 years, for today one can meet with enthusiasts who actually claim to have no interest in steam!

Along with Brian, George and Don, countless happy summer Saturdays were spent either at Worting Junction where the Bournemouth and west of England line diverged from Waterloo or on Basingstoke Station which lay a few miles towards London. Sometimes we camped at Worting Junction for a weekend and during the hot summer days would lie on the grassy banks listening to skylarks singing above and the glorious three-cylinder rhythms of the Bullied 'Pacifics'.

Traffic was invariably heavy and the scheduled expresses were augmented by holiday specials and freight traffic which produced 'N & U' Class 'Moguls' and 'S15' Class 4-6-0s.

Worting's rural charm contrasted with Basingstoke Station — a fine location for speed. Never will I forget the screaming whistles of the expresses as they approached and the way in which they roared through the station at speeds up to 90 mph, shaking the buildings and lifting newspapers from the bookstalls. The steam age's last great fling was being played out in epic manner and during those final years, speeds of 100 mph and more were recorded with increasing regularity by crews conscious of the traditions they were ending. Most of our winter visits to the Southern main were to

Basingstoke Station where we could be regularly fortified by mugs of hot coffee from the station buffet.

During June 1965 I walked the Bournemouth line westwards from Worting Junction to Micheldever, where the line runs through two 400-yard tunnels and I was invited by a Permanent Way ganger to accompany him through them. The two tunnels are separated by a 150-yard stretch of open track bounded by insurmountable rock cuttings on either side; here with his assistance I secured wonderful photographs of trains both traversing the short open stretch and entering the tunnel mouths. We were caught amidships in one of the tunnels by the 10.30 am ex-Waterloo travelling at high speed and headed by 'Merchant Navy' 35013, *Blue Funnel*. We dived into a protection cavity in the tunnel

One of the most remarkable survival stories in the history of the British steam locomotive concerned three of these Beattie 2-4-0 'Well Tanks' which remained at work on a lightly-laid China Clay line in Cornwall until 1962. The type was first introduced by the London and South Western Railway in 1864 for their suburban service from London Waterloo, but as this traffic grew they were rapidly displaced and the last example was withdrawn in 1898—apart from the three engines transferred to Cornwall! These were destined to outlive their sisters by 64 years and although considerably rebuilt over the intervening years, were finally withdrawn from service exactly one century after their original design was prepared.

wall just as she was entering from the eastern end, her screaming whistle resounded off the brickwork with earsplitting dimension as with terrifying noise and vibrations she screamed past us bouncing red hot cinders round our feet. The lights from the coaches combined with the swirling smoke threw ghostly shadows over the tunnel walls. Whilst waiting for the smoke to clear my spine tingled at the unmistakeable sound of another 'Pacific' approaching, this time on the up road. With whistle screaming and howling right through the tunnels 34104, *Bere Alston*, slammed past us with an express for Waterloo, emitting the thickest clouds of smoke imaginable. We waited some minutes for the smoke to clear before leaving the tunnel, whereupon the warm sunshine and grassy banks enabled me to recollect my senses. I thought I had seen the steam engine in all its moods until this experience in the Micheldever tunnels.

But my most lovely memories of the Southern were in the New Forest during the autumn of that year when, along with Judy Maddock I took a two week holiday in Boscombe. The weather was perfect and most days were spent in the forest next to the main line. The spells between trains were sometimes lengthy and here we could enjoy nature at its best. Small flowers abounded by the lineside, forest ponies were to be seen as fearless of man as they were of the trains, newts, lizards, snakes and toads became an everyday sight, not to mention the many species of butterflies, some of which are peculiar to the New Forest. Over many happy hours we picked blackberries and we would return to our guest house laden with them. For the entire fortnight, blackberry and apple pie graced the dinner menu to the delight of most residents.

Soon after this, Judy's parents moved from Leicester-shire to Poole and this gave us free access to the Bourne-mouth end of the line and many more thrilling weekends were spent until that final day in July 1967. The last year however lies outside the scope of this volume, as in 1966 I changed from monochrome to transparencies and the sad run down which culminated in that last traumatic weekend was recorded in colour. Many of the Southern engines which we had come to know so well, were despatched to Woodham's scrapyard in Barry and some still lie there to this day, though happily many have been preserved.

Below *An up stopping train heads through Winchfield behind BR 'Standard 4' No 75077 fitted with double-chimney. 80 of these 4-6-0s were built for general duties between 1951 and 1957.*

Right *The ugliest British steam locomotive of all time? This was the general consensus regarding Bulleid's 'Q1' Class, 40 of which were produced during World War 2. Grotesque they may be, but the 'Q1's were powerful and very free steaming and had the added distinction of being the last inside-cylinder 0-6-0 tender design to be built, thus ending a locomotive tradition of 110 years.*

Above *No one could have dreamt that Woodham's scrapyard in Barry would become little more than a repository for railway preservationists. Thousands of locomotives passed into private scrapyards during the 1960s and most disappeared within months, but despite Woodham's receiving some 250 engines, few were scrapped and this 'stay of execution' enabled preservationists to build up funds and save types which would otherwise have become extinct. In addition, dozens of individual engines have been rescued over the past 20 years. The centrepiece in this view of Woodham's sidings, is ex-Southern 'U' Class 'Mogul' No 31625 which arrived in June 1964. She lay in the scrapyard for 16 years, until purchased for preservation in 1980.*

Below *The pre-grouping atmosphere of Sway station makes a splendid foil for BR 'Standard 4' Class No 75076 as she whistles up to depart with a stopping train to Bournemouth.*

The 'King Arthurs' formed an important part of the Southern's express passenger fleet from the mid-1920s until the late 1950s, when some of their workings were taken over by the new BR 'Standard 5' 4-6-0s. Twenty of the new engines received names from the much loved 'King Arthurs' between 1959 and 1961 and here is No 73080, Merlin, *complete with 6ft 2in-diameter driving wheels.*

Waterloo

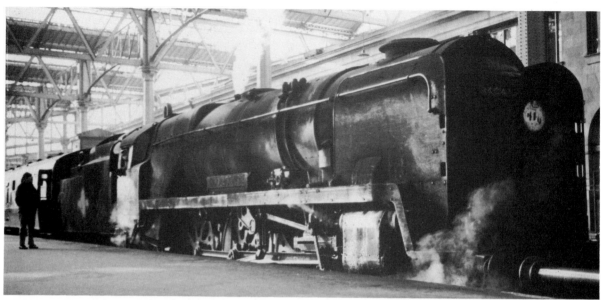

Left *We often commented that West Country 'Pacific' No 34044, Woolacombe, seemed exceptionally hard worked and frequently appeared twice in one day. Woolacombe had been a Bournemouth engine since the early '50s and possibly because of her high mileages she didn't quite make it to the end and was condemned in May 1967. Here we see* Woolacombe *against the buffer stops at Waterloo Station having arrived with an express from Southampton.*

Below left *A brace of Southern rebuilt West Country 'Pacifics' repose in the yard at Nine Elms depot London.*

On works

Below *These charming suburban tanks were introduced by Drummond for the London and South Western Railway in 1896. The class totalled 105 engines, but following the Southern's extensive electrification in the London area, these 'M7's spread to other parts of the system—especially for working branch lines. During their final years, many were relegated to shunting and the last ones could be seen on empty stock workings around Waterloo Station. All had disappeared by 1964, but No 30053—complete with 'Titfield Thunderbolt' chalked on her tank side—was overhauled and exported to America for preservation along with 'Schools' Class No 30926,* Repton, *(overleaf).*

Above *'Merchant Navy' Class No 35028,* Clan Line, *receives an intermediate overhaul at Eastleigh works prior to returning to main line service between London (Waterloo) Southampton, Bournemouth & Weymouth. Her shed plate, 70G, indicates that she is a Weymouth engine and along with six other 'Merchant Navys'* Clan Line *remained in service until finally ousted by electrification in July 1967.*

Below *Withdrawn from Basingstoke shed in December 1962, SR 'Schools' Class 4-4-0 No 30926,* Repton, *was stored until late 1966 when she was overhauled and restored to her former glory for subsequent despatch to the Steam Town Foundation, Vermont, Virginia USA.*

Isle of Wight

It is often said that time stands still on the Isle of Wight; certainly the pace of life there seems more akin to that of pre-war years. The island's railways mirrored this isolation, and until as recently as 1967 the beauty of a rural steam railway could be experienced. Motive power was provided by these lovely 0-4-4Ts introduced by Adams for the London and South Western Railway in 1889. Here, No W24 Calbourne—built in 1891 and transferred to the Isle of Wight in 1925—busily shunts coal for a local wharf whilst (below), sister engine No W14 heads a local passenger train.

Working junction

Left *The 'Bournemouth Belle' complete with immaculate rake of Pullman coaches heads westwards through Worting at high speed behind rebuilt 'West Country' No 34101 Hartland. The 'Belle' was one of the highspots of a day at the Southern main, for a superbly trimmed green 'Pacific' heading a train of chocolate and cream Pullmans was a wonderful sight to behold. The drab uniformity of our inter-city trains today makes poor comparison with such flamboyance.*

Below left *A freight from the Exeter line heads beneath the girder bridge carrying the up Bournemouth line at Worting Junction. Eastleigh had 20 of these useful BR 'Standard 3' 2-6-0s and they often appeared at Worting although we much preferred to see the more distinctive Southern types—'Q1's, 'N's, 'U's, or 'S15's—on freight.*

Below *Judy Maddock with No 34047, Callington. We all dreaded the day when the Southern would finish and when the final Sunday came, all remaining locomotives were despatched to Nine Elms, Weymouth or Salisbury for disposal. As we returned to the Midlands on that fateful evening, we stopped at Salisbury and found the shed crammed with engines—most of which had come in 'light' or coupled together in pairs that day. All fires had been dropped, but the engines were still in steam. It was an eerie and unforgettable experience for the shed was alive with the hissings and gurglings of the dying locomotives. Our emotions were supercharged by the tragedy and we continued our long journey home in silence.*

Next day, Judy wrote this poem which—despite technical imperfections—represents the feelings of a young girl who had only briefly known the steam age, and yet she captured the spirit of the time and in so doing, spoke for all of us . . .

A Last Tribute To Steam At Salisbury

The day is dark, the world seems grim
As I stroll through this once home, now tomb of steam,
So majestically poised, defiant of man
Accepting their fate with a graceful bow as would
only become a king.

Oh how pitiful stands my 'Dartmoor' and 'Bude', so
still,
Undressed of their well earned triumph, bountiful
glory, until
All that is left is their once loved frame,
To be shed to be scattered, to be put to shame.

Never again will these men of iron, men of fire,
of power, of speed,
Fly through our fields—crawl through our cities
Working so hard for love of their masters,
Fulfilling our every need.

Au'revoir my fine-loved friends the end for you
has come
And I must leave this once home of Salisbury,
my tribute to you is done.
But you will never be forgotten by our mortal men,
You will live in our hearts and minds for ever,
and ever, beyond our ken.

Left *The overbridge at Worting provided a splendid panoramic view of the four-track main line. It was an excellent vantage point for photography and here 'N' Class 'Mogul' No 31405 heads west with hoppers. The 'N's appeared in 1917 as a mixed-traffic design, but this engine was one of the 15 later examples built at Ashford works between 1932/4.*

Below left *That was a lovely weekend at Worting; the weather was glorious and we camped by the main for two nights. Apart from the wonderful trains, we had the timelessness of youth and good friends with which to enjoy it. It was on the Saturday morning that the 'S15' came through with a long mixed freight. We were very pleased to see her as at that time the 'S15s' were being rapidly withdrawn.*

New Forest

Below *A scene from our blackberry bridge deep in the New Forest with BR 'Standard 5' No 73155 at the head of a Bournemouth—London Waterloo express. No 73155 lasted until the end of Southern steam and was one of the engines despatched to Salisbury on the final Sunday and later included in the batch sent to Cashmores at Newport for breaking.*

Above *The stretch of line to the south of our blackberry bridge was perfect for photography or watching and we spent our entire autumn holiday in this area. Typical of the joys encounted is the view of an unidentified West Country 'Pacific' heading a London-bound express from Bournemouth. Note the third-rail in readiness for the impending electrification.*

Eastleigh

Left *A rare beast seen on a visit to Eastleigh in 1957 was No 30757 Earl of Mount Edgcumbe, one of two hefty 0-6-2Ts built in 1907 for the obscure and grandly titled Plymouth Devonport and South Western Junction Railway. Along with sister engine No 30758, Lord St Levan, Earl of Mount Edgcumbe was withdrawn from Plymouth shed in 1957 and despatched to Eastleigh for scrapping. However, as this scene indicates, she was retained as Eastleigh shed pilot for a while but she disappeared shortly afterwards, taking her class and her railway into history.*

Right *Shrouds of steam disperse to reveal the friendly face of Ivatt 2-6-2T No 41299 bursting from beneath the portals of Eastleigh shed. Though of LMS origin, considerable numbers of these general purpose engines were allocated to the Southern.*

Above *One of the great performers of World War 2 were these United States Army Transportation Corps 0-6-0Ts. The type saw widespread service in Europe and when hostilities ceased, many European railways absorbed the engines into their stocks. Some were stored at Newbury Racecourse until 1946 when the Southern Railway bought 14, Nos 30061–74, as their standard shunting type for Southampton Docks. They remained at Southampton until displaced by dieselisation in 1963. Some were scrapped; others passed to departmental service, whilst No 30067—seen here—became Eastleigh's works' pilot.*

Above right *Sunlight ripples through the decorative roof of Bournemouth Central Station and throws mottles over BR 'Standard 3' No 76059 as she waits to depart with a stopping train to Southampton.*

Right *During the Southern's last years, passengers between London, Southampton, Bournemouth and Weymouth became increasingly fascinated by the locomotives and the sights of a 'Merchant Navy' standing on the middle road at Bournemouth Central often attracted groups of bystanders. She is No 35013,* Blue Funnel, *one which survived until the end in July 1967. She was finally broken up by Buttigiegs of Newport in March 1968.*

Bournemouth

Above *The rebuilt 'Merchant Navys' were magnificent locomotives to behold and I think this study of No 35003,* Royal Mail, *reveals their charisma. She had just come on shed at Bournemouth having brought in an express from London Waterloo.* Royal Mail *survived until displaced by electrification in July 1967 and after being stored at Weymouth until November, was broken up by Cashmores in Newport the following month.*

Above right *The lovely air-smoothed casing of the un-rebuilt 'Light Pacifics' is shown to perfection in this late afternoon scene as No 34076, 41* Squadron, *leaves Bournemouth Central with an express for Waterloo. Aesthetically, I much preferred the unrebuilds despite their being irreverently referred to as 'Spam Cans'. Exactly 50 of the 'West Countries' and 'Battle of Britains' remained in their original condition out of the 110 built. However, all 30 of the larger 'Merchant Navys'* (**below**) *were rebuilt between 1956/9.*

Right *In contrast with its air-smoothed relation above, rebuilt 'Merchant Navy' No 35003,* Royal Mail, *makes a punctual departure at 5.13 to begin its 108-mile dash to London with the 'Bournemouth Belle'. This Pullman-car express was the most prestigious train of the day and the station always acquired an air of affluence before its departure.*

Chapter 7

The North-East

Fifteen years of train spotting were my apprenticeship to the making of the pictures in this volume and the following extract from my diary of December 20 1954, conveys something of the excitement experienced during those years, along with a glimpse of how dense railway operations were at the time. I was 14 when the extract was written and in an effort to capture the atmosphere the piece is reproduced in its original form. Over-use of the first personal pronoun will hopefully be excused. The piece is the last part of an account of a day trip to Leeds and begins as I left Farnley Junction shed...

'More than pleased, I slithered back down the steep bank and caught a bus to the centre of Leeds. The next shed was my first North Eastern one, Neville Hill 50B. I caught an 18 (Crossgates) tram and after another thrilling ride got off at Osmondthorpe Lane and walked down it. Presently I reached the shed (50B), the building is a very large one, I was hoping to see a 'D20' although 50B's two 'D20's had just gone to 50D. I did a little yard at the front of the shed and then entered the first large roundhouse and there, on the turntable, stood my first beautiful ex-North Eastern 'D20' No 62384 50D. It was clean and I got out my camera and even the black darkness did not stop me. I (perhaps without hope) took a photo and at the time of writing it has not yet been developed.

'I got more 'D49's and was pleased to get 'A3's No 60074 *Harvester*, No 60084 *Trigo* and No 60086 *Gainsborough*, all 50B. On the shed I saw several new classes; 4-6-2T '3P', 'A8' No 69882 (50B) (which are a Gresley rebuild of Raven Class 'D' 4-4-4T introduced 1913). Next of course the 'D20' already mentioned,

0-6-2T '3F', 'N13' No 69119 (50B) which was the only one on shed (the rest I saw in the yards), 0-4-4T '1P G5', Nos 67240, 67262, 67266, 67290, all 50B except No 67240 (50G). 0-8-0 '6F Q6' No 63436 (50B) (seen outside in the yards). Plenty of 'B16's were present and I left the shed and made my way to the goods yards where I could see 'N13's shunting.

'As I left the shed a terrific gust of wind almost knocked me over and with snow and rain stinging me in the face, I took refuge in a small brick building. At the time of my visit (according to my stock books) there were only five 'N13's in existence and I saw four of them, Nos 69119 (on shed) and shunting in the yards were Nos 69114, 69115 and 69117. The one I didn't get, No 69116 was scrapped a short time later so I thought it quite possible that it had already left for scrapping. The whole class is 50B. I took a photo also of a 'D49'. I copped two 'B1' namers Nos 61237 *Geoffrey H. Kitson* and No 61035 *Pronghorn*, both 50B. I had a total number of 37 cops on 50B. I now made my way back to the City Station where I was to meet the car at four o'clock. When I arrived at the station hall it was bang on four o'clock. As it happened the car did not come until about five o'clock and I for one was not sorry in the event of what I saw while waiting. After a look round I copped No 46437 (25C) and No 61053 (50A). It was at this moment I felt hungry as I had not had time for any dinner. So I had a snack in the refreshment room.

'I emerged feeling better and another look round produced more cops, this time a 'WD' 2-8-0 No 90684 (25G) and two more 'D49's both from 50D Starbeck Nos 62765 and 62749 *The Gothland* and the *Cottesmore*. Next was a 'B16' from 50C Selby No 61422.

My next cop was a beautiful sight it was L&Y 2-4-2T '2P' No 50795 (20E) which ran into the station in beautiful condition. It was the first one I had seen (not counting No 10897 at Uttoxeter). Tanker No 40147 20E was next. I walked up to the other end of the station where I copped No 61010 *Wildebeeste* (53B) waiting to take out the 5.08 pm to Hull. Standing near it was No 60080 *Dick Turpin* (52B) which was an excellent cop. I copped another 'G5' No 69270 (50B).

Then the car came and Leeds was soon left behind. We passed 25A Wakefield shed now lit up. On the way back I looked at *Locomotives of the Premier Line in pictures* which I had bought on the City Station. The next day I did my numbers and found I had a total of 132 cops'.

Darlington was one of Britain's lesser known railway towns and came to prominence when the North Eastern established its HQ and workshops there. To enthusiasts from the Midlands, Darlington was one of those far-away places where rarities could be found and the works was the site of an escapade which was typical of many at the time. It occured whilst we were cycling through the area on our way back from Scotland in 1955 and upon reaching Darlington works we found it was well secured and guarded. After surveying the periphery our best method of entry appeared to be over a wall. Railings carrying strands of barbed wire were built on top of the wall but by climbing onto our cycle crossbars we gained footholds in the crumbling brickwork and managed to prise the barbed wire up far enough to squeeze through,

This humble shunting design achieved a distinction without parallel in world locomotive history. Introduced by Worsdell for the North Eastern Railway, 85 of these 'J72's were built between 1898 & 1925—the last ones being under the LNER. Amazingly, a further 28 were built under British Railways between 1949 and 1951! Thus, the 'J72's were constructed over a 53-year period under three different stages of railway ownership and five regimes of locomotive superintendents!

whereupon we half jumped and half fell down into the works' yard the other side. We dashed between some lines of wagons and made our way towards the main shops. We were almost at the erecting shop when I noticed a pair of legs wearing black trousers on the other side of the wagons. Grasping Brian's arm I whispered 'Alarm Clock'. We ran the opposite way as quietly as we could, only to find another black-uniformed figure appear ahead; in desperation we turned to run the other way, only to find a third man; we were cut off.

As our captors advanced, one of the three was almost hysterical with rage. 'I saw you, I saw you', he shouted, 'I saw you climb under that wire. I saw you from my cabin!' We were taken to the security office and the police were called whilst one of the guards went to fetch our bikes. Then came the immortal lines we have chuckled about ever since:

'Do you walk into your neighbours garden';
'No' we mumbled.
'Do you walk behind the counter in Woolworths';
'No' we mumbled.
'THEN WHY DO YOU COME IN HERE?'
When the police arrived the guard continued his hysterical shouting, 'I saw them; I actually watched them get beneath that wire on yonder wall.' The police equally unenamoured by our escapades, took full detail of our names, families and reasons for being in Darlington, whereupon we were ordered never to set foot in the works again, before being ejected from the main gates with the final words, 'You will hear more about this in due course', ringing in our ears.

But we did go back, exactly ten years later, on a visit combined with our Scottish tour in 1965, hence the pictures you see here.

Flying Scotsman *was the most famous Gresley 'A3', but a contender for second place was No 60100* Spearmint. *For years she was the pride of Edinburgh (Haymarket) and featured prominently in railway folklore. Dieselisation of the East Coast main line caused* Spearmint *to be transferred to Edinburgh (St Margaret's) in 1962 where she remained until withdrawal in July 1965. She was despatched promptly to Darlington and breaking up proceeded the same month. This sad picture shows the thoroughbred partly dismantled but with nameplates still in position.*

Above *Coaling and watering at West Hartlepool depot for former NER 'Q6' 0-8-0 No 63407 resplendent in ex-works' condition in 1965! The 0-8-0 was the logical development of the 0-6-0 in the constant demand for more powerful locomotives; the full weight of the engine was available for adhesion and for slow heavy pulling over relatively straight networks it was ideal.*

Below *Darlington loco shed makes a fine backdrop to former LNER 'K1' Class 2-6-0 No 62012. The first 'K1' appeared in 1945 as a two-cylinder rebuild of a Gresley 'K4'. From 1949, 70 new 'K1's were built and although the type survived until 1967, they had a ridiculously short life for so small a general purpose engine.*

*The imposing stone portals of Tweedmouth shed make a fine frame for BR
'Standard' 2-6-0 No 77002. Twenty of these engines were built at Swindon in
1954—further orders being cancelled owing to the modernisation plan. The entire
class was allocated to Scotland and north-east England.*

A study in front ends celebrating the windshields and nameplate of LNER 'A1 Pacific' No 60154, Bon Accord.

Above *Shovelling hot ashes from the smokebox was one of the dirtiest jobs during routine servicing. Pictorially however, it was fascinating and here at West Hartlepool, a shed labourer clears an enormous deposit from the smokebox of a former NER Class 'Q6' 0-8-0.*

Above right *The definitive 'maid of all work' on Britain's railways for well over a century was the inside-cylinder 0-6-0 tender engine. Thousands of them embracing scores of different designs—all forged with the characteristics of their designers—ran the length and breadth of Britain for over a century.*

Right *The overall shape of the British steam locomotive changed dramatically following World War 2, as epitomised by these LMS Ivatt 2-6-0s introduced in 1947. The trend was towards two-cylinder designs with high running plates and many labour saving devices—including self-cleaning smokeboxes. These characteristics set the stage for BR's twelve standard designs from 1951 onwards.*

Left *Darlington works continued to overhaul steam locomotives until 1965 and one of the last engines they 'outshopped' was Stanier '8F' No 48100. Following the nationalisation of Britain's railways in 1948, 663 of these '8F's were absorbed and until 1964, they remained virtually intact. The type survived until the end of steam in August 1968.*

Below left *My visit to Darlington in 1965 was especially sad as this great works—which had built over 2,250 steam locomotives—was due to close early the following year. In the yard was the last ex-North Eastern 'J21' Class, No 65033 of 1886. The 'J21's were built at Darlington between 1886 and 1894 and were originally two-cylinder compounds. In the background is Stanier '8F' No 48689 along with 'A4' No 60010 Dominion of Canada—later restored at Crewe and exported to Montreal for preservation.*

Below *These magnificent engines were part of a succession of superb 0-8-0s used for heavy mineral hauling by the North Eastern Railway. No 63435 belonged to the 'Q6' Class which was comprised of 120 engines built between 1913 and 21. Extensively used on the north eastern coalfield—principally south of the Tyne—the 'Q6's remained intact until 1960 and the last ones were not withdrawn until 1967—one year before British steam ended.*

Above *Sunday afternoon at the engine sheds and nothing stirs; the only sounds are of dripping water and chattering sparrows. The depot's only human occupant is a shedman, but he is dozing in the quietness of the afternoon. But come six o'clock and the steamraisers will arrive to get these 'J27s' lit up. And then, during the early hours of Monday morning, they will leave the shed one by one to bring the Blyth coalfields alive once more.*

Above right *One sight at Neville Hill was as fine as any I had seen ten years earlier, for in the smoky depths of the roundhouse stood 'A1 Pacific' No 60118 Archibald Sturrock. The engine commemorated the first locomotive superintendent of the Great Northern Railway who held office between 1850 and 1866. I had seen No 60118 many times over the previous 15 years as she was a Leeds (Copley Hill) engine, and 'common through the main' at Grantham. She was one of the last survivors of the 50 'A1's and was not withdrawn until September 1965—five months after this picture was taken. She was finally broken up by Wards of Beighton, Sheffield, the following December.*

Right *The great partnership of steam and coal. And what better to epitomise it than a former North Eastern 'Q6' 0-8-0 at the head of a long drag.*

Above *Exactly ten years after my childhood escapades at Leeds (Neville Hill), I returned—this time better equipped for photography. Steam remained in evidence but the exotic diversity described earlier had gone. However, several 'A1's were present including No 60154, Bon Accord.*

Below *At Neville Hill shed was former Great Eastern 'N7' Class 0-6-2T No 69621, one of a class comprised of 134 engines introduced in 1915 for working the tightly-timed suburban trains—known as the 'Jazz Service'—from London Liverpool Street. She was the last GE locomotive to be built at Stratford works and upon withdrawal in 1962 was stored at Neville Hill pending preservation.*

Above *Another occupant of Neville Hill shed was No 3442, The Great Marquess—one of six 'K4' Class three-cylinder 2-6-0s designed by Gresley for the West Highland line from Glasgow—Fort William. She was built at Darlington in 1938 and upon withdrawal from Thornton in June 1961, was restored to original condition for preservation and stored temporarily at Neville Hill.*

Below *A stud of former North Eastern 'J27's bask amid the sooty depths of Blyth shed. These fine engines were the last in a long progression of inside-cylinder 0-6-0s for the NE and 115 were built at Darlington between 1906 and 1922. At the beginning of 1965 almost 50 remained in service.*

The crew water their steed in the form of ex-War Department 'Austerity' 2-8-0 No 90459 at West Hartlepool shed. No class in Britain's railway history was more neglected than these wartime 2-8-0s; they were always filthy and could be heard approaching from long distances by the clanging 'plonk' of banging bushes. For train spotters, they were the hardest of all British classes to 'clear', as 773 of them roamed Britain from dozens of different sheds and often on work which involved complex inter-colliery transfers and I never met anyone who had seen all of them. They were true workhorses, with a delightful habit of turning up miles off their home territory, but sometimes they were so grimy that their cabside numbers were impossible to read.

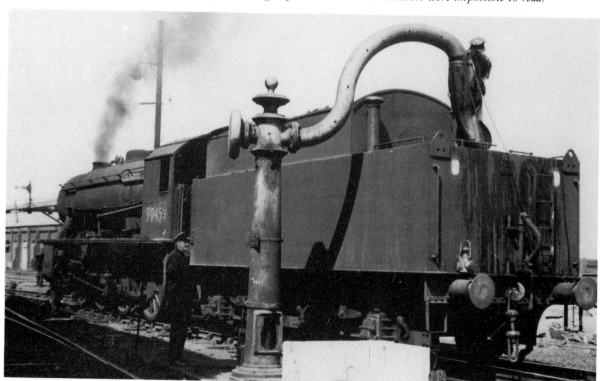

Chapter 8

Northamptonshire ironstone

Few people were interested in industrial railway networks during the golden years of British steam. Activities on the main lines were so engaging — both historically and aesthetically — that the industrial was largely ignored, despite there being thousands in existence in innumerable locations such as: collieries, iron and steel works, docks, power stations, gas works, mines, factories and even sewage works! The world's first steam locomotive was an industrial, as were all subsequent engines until 1825, albeit that these pristine creations bore little generic resemblance to the engines we know today.

The first true industrial appeared about 140 years ago in the form of a contractor's engine, used by the railway builders during the frenzied construction of the 1840/50s. These builders required a compact and economical engine, light enough to work over hastily-lain tracks, able to negotiate tight curves, yet having sufficient adhesion to draw heavy train loads of earth and materials.

From the outset, these engines polarised into very definite forms; they were either side or saddle tanks and either four or six coupled. Such engines found their way into every facet of industry, often spending their lives unseen by the populace at large. During a century of development, the industrial locomotive remained fundamentally unaltered either in shape or size.

Countless industrial railway networks existed throughout the country, many with up to 50 miles of track. In Leicester we were living within a few miles of one of the finest systems, for the Northamptonshire ironstone field was literally covered in industrial lines. The largest network by far consisted of innumerable lines radiating from Corby steelworks far into the surrounding countryside. But there were dozens of smaller concerns as well, many having connections with the Midland main line; their ore was conveyed to ironworks in the north as described in Chapter 3.

I developed an early awareness of industrials as a result of regular cycle journeys to Market Harborough, Kettering, Wellingborough and Northampton. The purpose of these trips was to visit the main sheds of those towns, but on the way I recall noting considerable ironstone activity and distinctly remember pacing two of Kettering Furnace's archaic 'Black Hawthorn' 0-4-0 STs storming along with a heavy train of loaded skips.

The Northamptonshire iron belt runs from central Lincolnshire across east Leicestershire, Rutland, Northamptonshire and into Oxfordshire. Smelting began in the area during the Iron Age and by Norman times the workings were believed to be extensive. For centuries, charcoal was used to smelt the ore, but so serious had the depletion of forests become that Parliament, during the reign of Elizabeth I, passed laws restricting the use of timber, and iron production in Northamptonshire ceased — the smelting being moved to Staffordshire and South Wales where an abundant supply of coal was to be found.

It was not until the construction of the railways that the vastness of the region's iron ore deposits was fully appreciated; the cuttings and tunnels revealed the presence of undreamt of quantities of ore. This set a stir among Victorian industrialists; samples of the ore were displayed at the Great Exhibition in 1851 and the rural areas of Northamptonshire were on the brink of a great boom period.

In 1852, Thomas Butlin of Wellingborough smelted the first pig iron of the new era. The railways had vastly increased the demand for iron whilst providing the perfect method of conveyance. The Midland railways' main line extension southwards from Leicester ran through the heart of the iron belt and numerous pits sprang up alongside between Market Harborough and Wellingborough. By 1872, over one million tons of iron-ore were being produced annually and Wellingborough became known as 'The Iron Town'; the night skies glowed with the new furnaces and it was predicted that Wellingborough would become a second Wolverhampton. However, the one missing element was coal and although numerous shafts were sunk it finally had to be admitted that the area had none.

In 1880, Samuel Lloyd of Lloyd & Lloyds the Birmingham tube makers, visited Corby and formed the Cardigan Iron Ore Company, having leased the mineral rights from Lord Cardigan, the landowner. This act was prompted by dwindling reserves of Staffordshire ore. Lloyds Ironstone was formed in 1885 to expand this operation and they pioneered a steam digger which revolutionised production. By 1910 the company had built two blast furnaces on a site close to the original quarries and the first Corby iron was in production. The Great War saw a vast increase in production, whilst massive extensions were undertaken during the 1930s which gave birth to Corby New Town to which vast numbers of workers, unemployed during the depression, gravitated from all parts of the country.

When the following pictures were made, Corby steelworks was at its height, employing 12,000 people with an annual capacity for steel tube of nearly one million tons. The subsequent decline of this great industry and tragic closure of the iron works in favour of the imported ore of higher iron content, into coastal steel works is widely known and today no iron ore is lifted from the Northamptonshire bed. Most of the mining areas have now been restored to agriculture but the region remains littered with earthworks — fragments of industrial archeology, the sole evidence of a once vibrant and thriving industry.

During the darkest days of World War 2, British hopes were centred on the long awaited invasion of Western Europe. New locomotives would be needed for this campaign and Edgar Alcock of Hunslet, convinced the Ministry of Supply that his design of powerful 0-6-0 'Austerity' saddle tank would be preferable to the LMS's 'Jinty' (page 121). The 'Austerities' played a major role in the campaign and spread to many countries. After the war, the engines were surplus to military requirements and passed into industry, especially the National Coal Board, who continued to order new examples until as late as 1964—by which time the class totalled 484 engines. The type also found favour in ironstone mines and remained in this service until the production of home ore ceased.

Nassington

The dying woodlands contrast sharply with Nassington's famous 'W170' Walking Dragline excavator. Open cast mining invariably involved removal of the overlying strata or 'overburden', and the 'W170' was able to remove four tons per bite and work to a depth of 65 ft. These draglines revolutionised ironstone mining and enabled deeper pits to be worked. However, when Nassington closed in 1971, the 'W170' was cut up on site.

By 1970, the tiny Northamptonshire village of Nassington had achieved two distinctions; firstly it was host to the last privately-owned ironstone mine and secondly, the mine operated the last two steam locomotives to work on British ironfields. Both were Hunslet 16 in 0-6-0STs named Jacks Green and Ring Haw respectively after two woods which lay within the company's 600 acres. Note the digger loading the wagons at the working face and the spent area behind. Note also the dynamited ore strata in the foreground and the heaps of silica white sand on top of the bank.

Corby Steelworks

Left *The locomotives in Corby steelworks were responsible for a wide range of duties; firstly they brought in the raw materials—limestone coal and slack—from the BR exchange sidings; secondly, they despatched the finished steel tubes and ingots; thirdly, they received the loaded ore trains from the mines division. Miscellaneous internal duties included hauling ladles of molten iron from the blast furnaces to the steel plant and the steel ingots thence to the rolling mills. Various furnace slags had to be conveyed; either to the asphalt plant for the manufacture of tar-macadam, or for use in the production of fertilisers. Other jobs included feeding coal to the huge coke ovens along with an enormous number of shunting duties.*

Below left and below *Dieselisation slowly encroached on Corby steelworks culminating with the arrival of five English Electric six-coupled diesel hydraulics. So ended the saga of Corby's yellow Hawthorn Leslies—a stud of engines which epitomised the industrial steam locomotive.*

The expansion of Corby steelworks in 1934, created a demand for more locomotives and six of these powerful 0-6-0STs were delivered from Hawthorn Leslie in that year. A further four arrived in 1936–38, followed by two more in 1940–41. They were the standard type for the steel division and during the early 1960s many were converted to burn Britoleum oil. These scenes, through the spectacle glasses of a sister engine, depict the tremendous activity in the loco shed yard.

Chapter 9

Crewe

There was no finer train watching place in the world than Crewe. Quite apart from having seven busy routes converging upon it and three loco sheds, the town was host to Britain's leading railway works; Crewe was without doubt Britain's principal railway town.

Crewe was but a village when the Grand Junction Railway built their works there in 1843. Three years later the Grand Junction joined with other railways to form the mighty London and North Western — or Premier Line as it was to be proudly known — and Crewe was adopted as the principal works for the whole network. The amazing total of 7,357 steam locomotives were built at Crewe, the last being '9F' Class 2-10-0 No 92250 in 1958. The works is believed to have effected 125,000 locomotive repairs between 1843 and 1967 (an average of 1,000 a year!) and at its peak 10,000 people were employed on the 147-acre site. Until as recently as 1940 the vast majority of Crewe's inhabitants were dependent in some way or another upon the railway.

By the age of 10, I was dreaming of Crewe and all the wonders it held and my first visit in 1951 was one of the happiest and most memorable days of my life. First we visited the north shed which provided passenger locomotives; line upon line of 'Patriots', 'Jubilees', 'Scots', 'Princess Royals' and 'Princess Coronations' along with innumerable smaller classes — an incredible array of motive power waiting to take passenger trains of every conceivable type bound for many parts of the country. But the works were an institution in themselves with 125 locomotives in every possible state of overhaul. The orderly discipline of a skilled workforce in this immacuately-run plant is something I

can never forget. It was a lesson and experience of inestimable value and promoted an understanding of railways which could never be learnt in university. Here were railways at their height in national service providing incomparable benefit to the country, for in 1951 railways were the pulse of the nation's economy. That such a vast national asset could crumble was as unimaginable to me as it would have been to the workforce. That such magnificence could ever be replaced by the puny motor trade run by selfish tom tiddlers is one of the incomprehensibilites of our age and one which will fascinate future historians. I say again, to have seen and witnessed everything that was Crewe in 1951 at the impressionable age of 11, was worth more than all the facts, figures and economics banded around by vapid academics intent upon the railways decline. As the poem written from the despatch of Crewe's last steam repair on February 2 1967 said in its closing lines:

'Great Britain played her greatest role
when locomotives ran on best steam coal.'

We made countless visits to Crewe, but seldom did we unofficially get round the works, so vast were they and so tight the security. Often we penetrated so far, and once, in earnest quest of the last LNWR coal tanks which had arrived for breaking up, we almost reached the scrap bay before being caught and ejected. The large south shed though not as prestigious as the north shed, was also difficult, and in between trying to get access to either the sheds or the works, we would spot from the over bridge at the north end of the station, where the activity was almost non-stop.

Crewe provided train spotters with all the thrill of the chase, for in addition to the works which overhauled innumerable classes from all over Britain, there was the enormity of regular line workings, not least on the freight side which brought rarities in by the hour. In fact, the closest thing I have found today to the spirit of Crewe is bird watching in Cley in North Norfolk during the autumn and spring migrations when apart from the normal prolific activities, virtually anything can — and does — turn up, for Cley is the Crewe of the ornithological world. As 75 different species can be recorded at Cley in one day, so could the same number of different classes be seen at Crewe.

Perhaps one brief story will illustrate the magic. By 1955, I had seen all of the 191 LMS 'Jubilees' except No 45645 *Collingwood*, — a Patricroft engine. This was no small accomplishment as the 'Jubilees' were distributed all over the country from Perth to Bristol and apart from being spread between 23 different depots were highly utilised and although all of them roamed beyond their regular territories from time to time, a tremendous amount of luck was needed to see them. For us in the Midlands, many of those allocated to the Central Division (Lancashire and Yorkshire) and Scotland were especially hard.

Despite innumerable Sunday visits to Patricroft, No 45645 was never 'on'; in fact of their six Jubilees, one would be lucky to find two at home, even on a Sunday. During one fruitless visit in 1957, I was told by the shed foreman that *Collingwood* had gone to Crewe works for overhaul. Overjoyed that I had at last run the engine to earth I scoured the railway press for visits to Crewe and found one on a Sunday three weeks hence. Excitedly I waited for the days to go by and as I passed through the works gate that afternoon I felt a surge of triumph. I estimated that my 'Jubilee' would have been on works for about a fortnight and so expected to see her in the erecting shop. The shops were full and

Left *After my triumph with Jubilee No 45645,* Collingwood, *at Crewe, I saw the engine on several subsequent occasions. One was at Patricroft and I had this picture taken to commemorate the conclusion of that eight year chase.* Collingwood *was finally condemned in November 1967; she proceeded directly to Crewe works and was broken up the same month.*

Right *Yard shunting at Crewe being ably performed by a former LMS 'Jinty' 0-6-0T. The engine is on the Shrewsbury line close to Crewe South shed.*

although I examined every frame and boiler the elusive *Collingwood* was missing. I became profoundly upset and tried to break away from the party to examine the erecting shop again, but was hustled on. But then, as we left the building and proceeded up the works' yard, there on a wagon stood No 45645's cab — not countable, but proof that my 'Jubilee' was there somewhere. How I missed the engine I will never know, but upon returning home that evening I wrote direct to Crewe requesting permission to join the next party to go round. Time passed and it was three weeks before I received permission to join a group in a fortnight's time. Five weeks I mused; I could be unlucky, my 'Jubilee' could be ex-shops by then.

The anguish of the wait can only be understood by those familiar with the passions of train spotting; all my friends and indeed everyone at the local lineside knew my dilemma and would be all too ready to laugh if *Collingwood* had returned to Patricroft.

I recall standing outside the main gate at two o'clock on the appointed day as our group waited for another party to come out; as they filed past I asked, 'Is No 45645 on?' No-one seemed to know until I snatched a notebook from one boy and avidly scanned the list and there to my surging joy, under the heading 'Paint Shop', was No 45645. It was difficult to contain my excitement as the group filed so slowly through the shops. Finally, after a long hour, we reached the door of the paint shop and I was the first inside; A 'Princess Royal', then two 'Scots' and beyond, yes — a 'Jubilee'; I ran up to it; '45645 *Collingwood*'. My last LMS named engine and one which gave me a 371 run in the stock book. An eight-year chase had ended.

Such was the magic of Crewe, playing its role in the railways' unparalled service to the nation; a service which incidentally — but not entirely insignificantly — gave boundless joy to a million boys as well.

Above and below *Crewe works employed the most amazing variety of locomotives for shunting and the conveyance of materials from shop to shop. Originally two gauges were in use; an inter-shop 18-in tramway and a huge network of standard gauge lines. On my first visit in 1951, the diversity included a pair of ancient L&NWR 0-4-2 pannier tanks; we called them 'Bissels' and I greatly appreciated seeing them. By the mid-60s however, all operations within the works were handled by 'Jintys'.*

Below right *The magic that was Crewe. In the huge erecting shop alone, over 50 locomotives could be seen in varying stages of overhaul. 'Britannia Pacific' No 70022 Tornado undergoes her last shopping in 1965. Withdrawn from Carlisle (Kingmoor) shed in December 1967, Tornado was broken up by Wards of Inverkeithing in April the following year.*

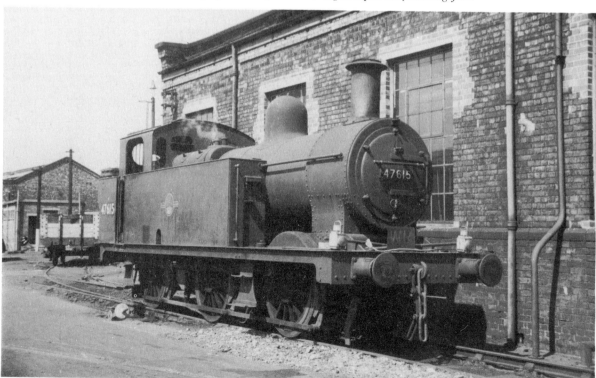

Above *Britain's ultimate express passenger engine and evolution's last fling. The solitary and now legendary 'Duke' Class No 71000, Duke of Gloucester, stored complete with bagged chimney outside the paint shop at Crewe works. Built in 1954 with three cylinders and Caprotti valve gear, No 71000 was intended to be the prototype of a new standard design for the fastest and heaviest expresses, but the modernisation plan of the following year caused the programme to be abandoned. Withdrawn in November 1962, No 71000 was stored at Crewe for five years pending possible preservation until despatched to Barry Docks for scrapping. She immediately became the subject of a private preservation bid which proved successful and in 1974, she left Barry and is currently being restored to full working order.*

Above *The sensuous form of the 'Jubilee'—complete with 6ft 9in-diameter driving wheels—is caught in this study of the 'class leader,' No 45552, Silver Jubilee. Note the raised numerals on the cabside. In 1952, I 'copped' Silver Jubilee on Crewe (South) and my shouts of joy aroused the shed foreman and we got 'caught'. This picture of her at Crewe (North), was made exactly 12 years later in 1964, shortly before this celebrated engine was condemned. She was broken up by Cashmores at Great Bridge in 1965.*

Right *Days beyond recall in the pleasant company of British Railway's '9F' 2-10-0 No 92083; standing in the cab is Judy Maddock, with Brian Stafford (front left) and Derek Yoxen.*

Chapter 10

Lancashire — steam's last outpost

My first visit to Lancashire during the early 1950s enabled me to see the industrial north as it was in the post-war years, before the intensive clean up campaign, — combined with industrial decline — took effect. As a Briton I am proud to have had this experience for I saw with young eyes those blackened remains of the workshop of the world and I saw them laced in the misty grime-laden atmosphere of a Lowry painting. Such a landscape was as dramatic as any on earth, yet so little was ever recorded visually — industrial subjects being eschewed in art. Today, no comparable industrial tapestry can be found and only odd selected view points

These Fowler 'Parallels' were an everyday sight in the Midlands. They worked the Leicester to Nottingham, Burton and Birmingham locals and often on our day trips to Rugby, we would be hauled by one of Leicester's two examples, Nos 42330/1. Introduced in 1927, the class totalled 125 engines and they were the forerunners of the Stanier, Fairburn and BR 'Standard' 2-6-4Ts (page 33 and page 18). Scrapping commenced in 1959 and No 42343—seen at Buxton during the mid-60s—was one of the last survivors.

around the world can capture a flavour of what it was like. But in exactly the same way that the derelict L&Y at Durran Hill first kindled my passion for derelicts, those childhood impressions of Lancashire engendered a deep passion for the industrial landscape's stark beauty.

And how the names of Lancashire's great locomotive sheds will live on amid industrial legend; Newton Heath, Gorton, Longsight, Patricroft, Trafford Park and Springs Branch. And the great Manchester stations — cathedrals to the age of steam — Central, Exchange, London Road, Victoria. Few places on earth have a greater railway heritage than Manchester for from that city came in addition the famous railway builders; Vulcan Foundry at Newton-le-Willows, Beyer Peacock at Gorton, whilst only a few miles to the west beyond the Manchester Ship Canal lay the Patricroft works of Nasmyth Wilson.

The changes wrought over the last 15 years of steam have already been touched upon, but at least Britain's last steam locomotives worked from Lancashire sheds. They lingered on until August 1968, one year after the Southern had finished. Enthusiasts from all over Britain flocked to Lancashire to pay their last respects to the grimy Stanier 'Black 5's and '8F's eking out their days at the North Lancashire depots of Rose Grove, Lostock Hall and Carnforth. Watching these sheds slowly fade into industrial legend was harrowing and I used to think back to that morning in 1955 when my father showed me the headlines in the national papers announcing the modernisation plan for railways, under which steam traction was to be phased out. But steam engines were a part of the world — an institution as permanent as the streets upon which we walked — and they certainly wouldn't disappear just because the government had drawn up a plan. Governments were always drawing up plans and even if they did carry it through it was bound to take ten times longer than they said. Everyone at the lineside agreed, so we didn't worry about it.

Such was the faith of adolescence, but before many years had passed Britain's steam heritage and indeed more than half of the railway itself was destroyed. The plan had been implemented far more viciously than seemed credible and with it the most awe-inspiring of man's creations disappeared from our homelands.

When British Railways finally condemned its last steam locomotives on August 9 1968, the list included several Stanier '8F' 2-8-0s. They had eked out their last months in begrimed condition and with a minimum of maintenance. It was hard to believe that only ten years earlier, almost 700 of them had been employed on heavy main line freight hauls throughout many parts of Britain.

Left *Former Midland '4F' No 43893 was a regular Skipton engine over many years of her working life and is depicted fully equipped for snowplough duty over the Settle and Carlisle line. I always admired the '4F's and never failed to be impressed by their ability to lift heavy freights and keep them rolling. No 43893 belonged to a batch of 192 engines built by the Midland between 1911 and 1922. The type was perpetuated by the LMS and building continued until 1940, when a grand total of 772 engines had been reached.*

Below left *One of the most familiar of the 60 'Jubilees' which passed regularly beneath my spotting bridge at Newton Harcourt was No 45562 Alberta. She was a Leeds (Holbeck) engine and a regular performer on the St Pancras—Leeds—Bradford expresses. Alberta was one of the last two 'Jubilees' to remain in service—the other being No 45593, Kolhapur. Both were withdrawn from Holbeck in October 1967; Kolhapur went for preservation, whilst Alberta was despatched to Cashmores at Great Bridge for scrapping.*

Lancashire landscape

Below *A foggy Sunday afternoon at Newton Heath shed in Manchester with the pilot engine slipping vigorously on the damp rails as she places dead engines into position for the steam raiser, in readiness for the new week's diagrams.*

Amid the stark Lancashire landscape of a Lowry painting, a brace of former
Lancashire and Yorkshire 0-6-0s stand in the shed yard at Newton Heath.
Lowry—who came from nearby Salford—epitomised scenes such as this in his
paintings before the far reaching economic and environmental changes of the late
1950s swept away the incomparable landscape of the industrial north.

Below *From an early age, I learnt to treat rumours about locomotives with a healthy suspicion and when I was told that an ancient Lancashire and Yorkshire Railway Barton Wright 0-4-4T was hidden amid Cheetham Hill carriage sidings in Manchester on stationary boiler duties I dismissed it. Barton Wright's tenure of office had ceased in 1886 and the last of his little known 0-4-4Ts had been withdrawn in 1921. But the rumours persisted, until one foggy day when we went to explore the huge sidings at Cheetham Hill and there—to our amazement—stood this incredible survivor from a lost dynasty.*

Right *Few classes were more exciting to spot than the former LMS 'Crab' 2-6-0s. 245 were put into service between 1926 and 1932. In 1950, the 'Crabs' were distributed between more than 40 depots ranging from Perth to London and their complex freight diagrams and propensity to work specials, extras and excursions, meant that every one seen was a potential rarity. Somehow, their shape exuded rarity; capuchon chimney, huge inclined cylinders, curved running plate and a tantalising aura of Lancashire and Yorkshire Railway aesthetics.*

Below right *'Crab' on a mixed freight; No 42900 heads northwards through Stockport. These fine engines once hauled the type of freight now handled by juggernauts on motorways. From express perishables to pigeon specials, the 'Crabs' brought home the goods safely, efficiently, reliably, kindly to the environment and above all, economically!*

The Crabs

Index to locomotive types

Ex-GN 'J50'	0-6-0T	38, 39, 40, 41
Ex-NE 'J72'	0-6-0T	97
Ex-NE 'N13'	0-6-2T	96
Ex-NB 'N15'	0-6-2T	29
Ex-GE 'N7'	0-6-2T	108
Ex-NE 'A8'	4-6-2T	96
Ex-WD 'Austerity'	2-8-0	35, 42, 43, 96, 110

British Railways

'Britannia' Class	4-6-2	30, 31, 34, 122, 123
'71000' Class	4-6-2	123
'Clan' Class	4-6-2	30, 31
'73000' Class	4-6-0	19, 30, 44, 76, 81, 89
'75000' Class	4-6-0	78, 80
'76000' Class	2-6-0	86, 87, 92, 93
'77000' Class	2-6-0	100
'78000' Class	2-6-0	46, 47, 48, 49, 58, 59
'80000' Class	2-6-4T	18, 126
'92000' Class	2-10-0	35, 40, 41, 50, 51, 119, 124, 125
English Electric 1750 HP Diesel		66

Industrial

'Black Hawthorn'	0-6-0ST	111
'Hawthorn Leslie'	0-6-0ST	116, 117, 118
Hunslet 16-in	0-6-0ST	114, 115
Hunslet 'Austerity'	0-6-0ST	112
English Electric 'DH'		116